Springer Series in Optical Sciences   Volume 70

Springer
Berlin
Heidelberg
New York
Barcelona
Hong Kong
London
Milan
Paris
Singapore
Tokyo

Springer Series in Optical Sciences

Editorial Board: A. L. Schawlow A. E. Siegman T. Tamir

Managing Editor: H. K. V. Lotsch

42 **Principles of Phase Conjugation**
By B. Ya. Zel'dovich, N. F. Pilipetsky,
and V. V. Shkunov

43 **X-Ray Microscopy**
Editors: G. Schmahl and D. Rudolph

44 **Introduction to Laser Physics**
By K. Shimoda 2nd Edition

45 **Scanning Electron Microscopy**
Physics of Image Formation and Microanalysis
By L. Reimer 2nd Edition

46 **Holography and Deformation Analysis**
By W. Schumann, J.-P. Zürcher, and D. Cuche

47 **Tunable Solid State Lasers**
Editors: P. Hammerling, A. B. Budgor,
and A. Pinto

48 **Integrated Optics**
Editors: H. P. Nolting and R. Ulrich

49 **Laser Spectroscopy VII**
Editors: T. W. Hänsch and Y. R. Shen

50 **Laser-Induced Dynamic Gratings**
By H. J. Eichler, P. Günter, and D. W. Pohl

51 **Tunable Solid State Lasers for Remote
Sensing** Editors: R. L. Byer, E. K. Gustafson,
and R. Trebino

52 **Tunable Solid-State Lasers II**
Editors: A. B. Budgor, L. Esterowitz,
and L. G. DeShazer

53 **The CO_2 Laser** By W. J. Witteman

54 **Lasers, Spectroscopy and New Ideas**
A Tribute to Arthur L. Schawlow
Editors: W. M. Yen and M. D. Levenson

55 **Laser Spectroscopy VIII**
Editors: W. Persson and S. Svanberg

56 **X-Ray Microscopy II**
Editors: D. Sayre, M. Howells, J. Kirz,
and H. Rarback

57 **Single-Mode Fibers** Fundamentals
By E.-G. Neumann

58 **Photoacoustic and Photothermal Phenomena**
Editors: P. Hess and J. Pelzl

59 **Photorefractive Crystals
in Coherent Optical Systems**
By M. P. Petrov, S. I. Stepanov,
and A. V. Khomenko

60 **Holographic Interferometry
in Experimental Mechanics**
By Yu. I. Ostrovsky, V. P. Shchepinov,
and V. V. Yakovlev

61 **Millimetre and Submillimetre Wavelength
Lasers** A Handbook of cw Measurements
By N. G. Douglas

62 **Photoacoustic
and Photothermal Phenomena II**
Editors: J. C. Murphy, J. W. Maclachlan Spicer,
L. C. Aamodt, and B. S. H. Royce

63 **Electron Energy Loss Spectrometers**
The Technology of High Performance
By H. Ibach

64 **Handbook of Nonlinear Optical Crystals**
By V. G. Dmitriev, G. G. Gurzadyan,
and D. N. Nikogosyan 3rd Edition

65 **High-Power Dye Lasers**
Editor: F. J. Duarte

66 **Silver Halide Recording Materials
for Holography and Their Processing**
By H. I. Bjelkhagen 2nd Edition

67 **X-Ray Microscopy III**
Editors: A. G. Michette, G. R. Morrison,
and C. J. Buckley

68 **Holographic Interferometry**
Principles and Methods
Editor: P. K. Rastogi

69 **Photoacoustic
and Photothermal Phenomena III**
Editor: D. Bićanic

70 **Electron Holography**
By A. Tonomura 2nd Edition

71 **Energy-Filtering Transmission
Electron Microscopy**
Editor: L. Reimer

Volumes 1–41 are listed at the end of the book

Akira Tonomura

Electron Holography

Second, Enlarged Edition
With 127 Figures

Springer

Dr. AKIRA TONOMURA
Advanced Research Laboratory
Hitachi, Ltd.
Hatoyama-cho, Hiki-gun
Saitama 350-03, Japan

Editorial Board

ARTHUR L. SCHAWLOW, Ph. D.
Department of Physics, Stanford University
Stanford, CA 94305-4060, USA

THEODOR TAMIR, Ph. D.
Polytechnic University
333 Jay Street, Brooklyn, NY 11201, USA

Professor ANTHONY E. SIEGMAN, Ph. D.
Electrical Engineering
E. L. Ginzton Laboratory, Stanford University
Stanford, CA 94305-4060, USA

Managing Editor: Dr.-Ing. HELMUT K.V. LOTSCH
Springer-Verlag, Tiergartenstrasse 17, D-69121 Heidelberg, Germany

ISSN 0342-4111
ISBN 3-540-64555-1 2nd Edition Springer-Verlag Berlin Heidelberg New York
ISBN 3-540-57109-4 1st Edition Springer-Verlag Berlin Heidelberg New York

Library of Congress Cataloging-in-Publication Data

Tonomura, A. (Akira), 1942– Electron holography / A. Tonomura. – 2nd, rev. ed. p. cm. – (Springer series in optical sciences ; v. 70) Includes bibliographical references and index. ISBN 3-540-64555-1 (alk. paper) 1. Electron holography. I. Title. II. Series. QC449.3.T66 1999 502'.8'25–dc21 99-10556 CIP

This work is subject to copyright. All rights are reserved, whether the whole or part of the material is concerned, specifically the rights of translation, reprinting, reuse of illustrations, recitation, broadcasting, reproduction on microfilm or in any other way, and storage in data banks. Duplication of this publication or parts thereof is permitted only under the provisions of the German Copyright Law of September 9, 1965, in its current version, and permission for use must always be obtained from Springer-Verlag. Violations are liable for prosecution under the German Copyright Law.

© Springer-Verlag Berlin Heidelberg 1993, 1999
Printed in Germany

The use of general descriptive names, registered names, trademarks, etc. in this publication does not imply, even in the absence of a specific statement, that such names are exempt from the relevant protective laws and regulations and therefore free for general use.

Typesetting: PS™ Technical Word Processor
Cover concept by eStudio Calamar Steinen using a background picture from The Optics Project. Courtesy of John T. Foley, Professor, Department of Physics and Astronomy, Mississippi State University, USA
Cover production: *design & production* GmbH, Heidelberg
SPIN 10680006   57/3144/di-5 4 3 2 1 0 - Printed on acid-free paper

Preface

Electron holography, a two-step imaging method taking advantage of light and electrons, has been employed for fundamental experiments in physics as well as for high-precision measurements in microscopy. It continues to be advanced in its applications though the speed of the development is not high but steady together with technical improvements of both the coherent electron beam and the reconstruction system. Progress has been made since the publication of the first edition of this book in 1993, and this second edition with new results integrated is herewith made available.

I hope that the present monograph will help develop this interesting field of electron interference further.

Hatoyama *Akira Tonomura*
December 1998

Preface to the First Edition

Electron holography has recently paved a new way for observing and measuring microscopic objects and fields that were previously inaccessible employing other techniques. Full use is made of the extremely short wavelength of electrons, enabling electron holography to have a great impact on fields ranging from basic science to industrial applications. This book will provide an overview of the present state of electron holography for scientists and engineers entering the field. The principles, techniques and applications which have already been developed, as well as those which are expected to arise in the near future, will be discussed. The strange and interesting nature of electron quantum interference has intrigued the author who has devoted most of his life to this field, and motivated him to write this book in the hope of raising interest in, and encouraging others to enter, the field.

The author would like to express his sincere thanks to Professor Emeritus Koichi Shimoda of Tokyo University, his mentor during his university days. His sincere gratitude goes to Professor Chen Ning Yang of SUNY Stony Brook, Professor Hiroshi Ezawa of Gakushuin University, Professor Emeritus Ryoji Uyeda of Nagoya University, and Dr. Akira Fukuhara of Hitachi, Ltd. for their continued encouragement and helpful suggestions regarding the electron-holography experiments. Thanks are also due to the following colleagues for their collaboration in carrying out electron holography experiments: Tsuyoshi Matsuda, Dr. Junji Endo, Dr. Nobuyuki Osakabe, Dr. Takeshi Kawasaki, Dr. Takao Matsumoto, Takaho Yoshida, Dr. Ken Harada, Dr. John Bonevich, Hiroto Kasai, Shigeo Kubota, and Shokichi Matsunami of Hitachi, Ltd., Professor Giulio Pozzi of Lecce University, and to Dr. Kazuo Ishizuka, Dr. Takayoshi Tanji, Dr. Quingxin Ru, and other members of the Research Development Corporation of Japan (JRDC). Special thanks are due to Mari Saito and Yuka Sugao of JRDC and Hiromi Yamasaki of Hitachi, Ltd. for typing the manuscript and preparing the figures. Lastly, my gratitude goes to Dr. Koichi Urabe of Hitachi, Ltd. for reading and correcting the manuscript.

Hatoyama *Akira Tonomura*
August 1993

Contents

1. Introduction 1
2. Principles of Holography 2
   2.1 In-Line Holography 3
   2.2 Off-Axis Holography 6
   2.3 Holography Using Two Different Kinds of Waves 8
3. Electron Optics 10
   3.1 Electron Microscopy 10
      3.1.1 Ray Diagram 10
      3.1.2 Electron Guns 10
      3.1.3 Electron Lenses 12
   3.2 Interference Electron Microscope 13
   3.3 Coherence Properties of Electron Beams 15
      3.3.1 Temporal Coherence 16
      3.3.2 Spatial Coherence 17
4. Historical Development of Electron Holography 20
   4.1 In-Line Holography 21
   4.2 Off-Axis Holography 25
5. Electron Holography 29
   5.1 Electron-Hologram Formation 29
      5.1.1 Ray Diagram 29
      5.1.2 Experimental Apparatus 31
         a) Electron Gun and Illumination System 31
         b) Electron Interferometer 32
         c) Recording System 33
   5.2 Image Reconstruction 34
      5.2.1 Interference Microscopy 34
      5.2.2 Phase-Amplified Interference Microscopy 36
         a) Optical Method 36
         b) Numerical Method 39
         c) Phase-Shifting Method in Optical Reconstruction 39
         d) Phase-Shifting Method in Electron Microscopy 40
      5.2.3 Three-Dimensional Imaging Reconstruction 41
      5.2.4 Real-Time Observations 42
      5.2.5 Image Restoration by Aberration Compensation 44
      5.2.6 Micro-Area Electron Diffraction 47

6. Aharonov-Bohm Effect: The Principle Behind the Interaction of Electrons with Electromagnetic Fields ... 50
6.1 What is the Aharonov-Bohm Effect? ... 51
6.2 Unusual Features of the Aharonov-Bohm Effect: Modified Double-Slit Experiments ... 53
6.3 The History of Vector Potentials ... 55
6.4 Fiber-Bundle Description of the Aharonov-Bohm Effect ... 56
6.5 Early Experiments and Controversy ... 60
   6.5.1 Early Experiments ... 60
   6.5.2 Nonexistence of the Aharonov-Bohm Effect ... 61
      a) Non-Stokesian Vector Potential ... 61
      b) Hydrodynamical Formulation ... 62
      c) Doubts About the Validity of Early Experiments ... 63
   6.5.3 Dispute About the Nonexistence of the Aharonov-Bohm Effect ... 63
      a) Non-Stokesian Vector Potentials ... 63
      b) Hydrodynamical Formulation ... 64
      c) Discussions on the Validity of Experiments ... 65
6.6 Experiments Confirming the Aharonov-Bohm Effect ... 66
   6.6.1 An Experiment Using Transparent Toroidal Magnets ... 66
      a) Sample Preparation ... 66
      b) Experimental Results ... 68
      c) Discussions of the Validity of the Experiment ... 69
   6.6.2 An Experiment Using Toroidal Magnets Covered with a Superconducting Film ... 70
      a) Sample Preparation ... 70
      b) Experimental Results ... 72

7. Electron-Holographic Interferometry ... 78
7.1 Thickness Measurements ... 78
   7.1.1 Principle of the Measurement ... 78
   7.1.2 Examples of Thickness Measurement ... 79
   7.1.3 Other Applications ... 82
7.2 Surface Topography ... 83
7.3 Electric Field Distribution ... 84
7.4 Domain Structures in Ferromagnetic Thin Films ... 85
   7.4.1 Measurement Principles ... 85
   7.4.2 Magnetic Domain Walls in Thin Films ... 87
7.5 Domain Structures in Fine Ferromagnetic Particles ... 90
7.6 Magnetic Devices ... 93
7.7 Domain Structures in Three-Dimensional Particles ... 96
7.8 Three-Dimensional Image ... 99
   7.8.1 Electric Potentials ... 99
   7.8.2 Magnetic Fields ... 100
7.9 Dynamic Observation of Domain Structures ... 102

7.10	Static Observation of Fluxons in the Profile Mode	103
	7.10.1 Quantized Flux (Fluxons)	103
	7.10.2 Experimental Method	104
	7.10.3 Experimental Results	105
7.11	Dynamic Observation of Fluxons in the Profile Mode	108
	7.11.1 Thermally Excited Fluxons	108
	7.11.2 Current-Driven Fluxons	110
	a) Experimental Method	111
	b) Experimental Results	116
7.12	Observation of Fluxons in the Transmission Mode	117
	7.12.1 Experimental Methods	117
	7.12.2 Experimental Results	118
	a) Behavior of Fluxons in Nb Thin Films	118
	b) Estimation of Pinning Forces of Defects	121
	c) Intermittent Rivers of Fluxons	122
	d) Matching Effect	124
	e) High-T_c Superconductors	130

8. High-Resolution Microscopy ... 133
8.1 Phase Contrast Due to Aberration and Defocusing ... 133
8.2 Optical Correction of Spherical Aberration ... 138
  8.2.1 In-Focus Electron Micrograph of a Crystalline Particle ... 138
  8.2.2 Off-Axis Hologram of a Crystalline Particle ... 139
  8.2.3 Image Reconstruction ... 141
  8.2.4 Spherical Aberration Correction ... 141
8.3 Numerical Correction of Spherical Aberration ... 143

9. Conclusions ... 146

References ... 147

Subject Index ... 159

1. Introduction

Electron holography is an imaging technique that records the electron interference pattern of an object on film (hologram) and then reconstructs an optical image by illuminating the hologram with a laser beam. In this process, electron wave fronts are transformed into optical wave fronts. Images of microscopic objects and fields that are so small that they can be observed only by using an electron beam with an extremely short wavelength are enlarged and reconstructed on an optical bench. This allows versatile optical techniques to be applied to overcome the limitations of electron microscopes.

Holography is now widely known − not only by scientists but also by artists, philosophers, and the general public − as a kind of stereoscopic photography using a laser beam. Holography was, however, originally invented in 1949 by *Dennis Gabor*, as a way of breaking through the resolution limit of electron microscopes [1.1, 2]. The resolution of electron microscopy is not determined by the fundamental limitation, the electron wavelength, but by the large aberrations of the objective lens. It is impossible to construct an aberration-free lens system by combining convex and concave lenses due to the lack of any practical concave lens. *Gabor* intended to optically compensate for the aberrations in the reconstruction stage of holography. The intrinsic value of holography was not fully recognized until 1962, when *Leith* and *Upatnieks* [1.3] reconstructed clear images by utilizing a coherent laser beam. Similarly, practical applications of electron holography have been made possible by the development of a coherent field-emission electron beam [1.4]. With this beam, electron holography has made a remarkable step towards new and practical applications [1.5-14].

To take a concrete example, the phase distribution of the electron wave function transmitted through an object has become observable to within a measurement precision as high as $2\pi/100$, while in electron microscopy only the intensity distribution can be observed. This technique has enabled us to directly observe magnetic lines of force of a magnetic object: The contour fringes in the interference micrograph follow magnetic lines in h/e ($=4 \cdot 10^{-15}$ Wb) units. This was applied to actual problems such as the magnetic-domain structure of a ferromagnetic thin film and also to the observation of magnetic fluxons penetrating a superconductor.

2. Principles of Holography

Holography, being a unique imaging technique that does not use lenses, is based on the most fundamental properties of waves, interference and diffraction. Holography is therefore applicable to all kinds of waves – light, X-ray, sound, electron, or neutron waves – regardless of whether there is a lens involved for the wave. The major feature of holography is that a complete wave (i.e., a complex amplitude) can be reconstructed from an exposed film called a **hologram** (a photograph containing all information, amplitude and phase). For this reason, laser holography can produce a far more realistic stereoscopic image than can be provided by any other technique.

The first step of holography consists of recording an interference pattern, or *hologram*, between the reference wave ϕ_r and the wave ϕ_0 scattered from an object. Here, ϕ_0 and ϕ_r represent the complex amplitudes of these waves. Since these two waves are partial waves emitted from a common source, they are coherently superposed at the hologram plane to interfere with each other. The intensity I of the interference pattern is given by

$$I = |\phi_0 + \phi_r|^2 . \tag{2.1}$$

When the interference pattern is exposed onto a film and developed, the amplitude transmittance t of the film is given by

$$t = I^{-\gamma/2} = |\phi_0 + \phi_r|^{-\gamma} , \tag{2.2}$$

where γ indicates the **contrast value** of the film. If the fillm is reversed and the contrast is inverted to $\gamma = -2$, t becomes proportional to I, though this condition is not always necessary for image reconstruction.

The second step in holography consists of reconstructing the image of the original object. For simplicity, the hologram is illuminated with the same reference wave as that used in forming the hologram. The transmitted amplitude T is then given by

$$T = |\phi_0 + \phi_r|^2 \phi_r = (|\phi_0|^2 + |\phi_r|^2)\phi_r + \phi_0|\phi_r|^2 + \phi_0^* \phi_r^2 . \tag{2.3}$$

The imaging properties of holography can be understood simply by interpreting the terms in this equation. The first term corresponds to the transmitted wave, and the second to the wave scattered from the object. This means that an exact image can be reconstructed if the second term can be observed independently of the others. The third term is similar to the second one, but its phase value is opposite in sign. It produces a conjugate image, the amplitude of which is the complex-conjugate of the reconstructed image.

The image formation is essentially the same, even when the hologram is illuminated with a wave whose wavelength differs from that of the original reference wave. Parameters such as the image magnification and the distance between the hologram and the image depend on the ratio of the two wavelengths. This will be discussed in more detail in Sect. 2.3. Up to this point, holographic image formation seems to work perfectly, but this is not the case when higher-order terms are taken into consideration. The aberrations associated with this type of imaging are similar to those for imaging with an optical lens [2.1].

2.1 In-Line Holography

The simplest way of producing a hologram is illustrated in Fig. 2.1a which shows a point object illuminated with a plane wave. The transmitted plane wave plays here the role of a reference wave. This type of holography is called **in-line holography** because the object and reference waves propagate along a line. The amplitudes of the reference and object waves can be expressed as

$$\phi_r = e^{ikz} \quad \text{and} \quad \phi_0 = i\frac{f}{r}e^{ikr}, \tag{2.4}$$

where f is the scattering amplitude from the point object. Then, the intensity distribution $I(x,y)$ in the hologram plane at a distance l from the object is given by

$$I(x,y) = |\phi_0 + \phi_r|^2 \approx 1 + \left(\frac{f}{l}\right)^2 - \frac{2f}{l}\sin\left(\frac{k(x^2+y^2)}{2l}\right), \tag{2.5}$$

if $l^2 \gg x^2 + y^2$ and consequently $r = \sqrt{l^2+x^2+y^2} \approx l + (x^2+y^2)/(2l)$.

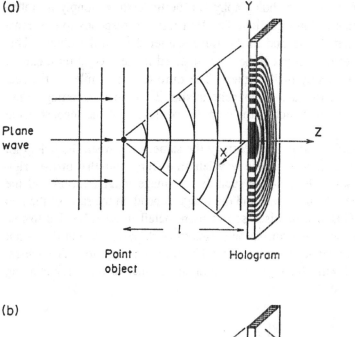

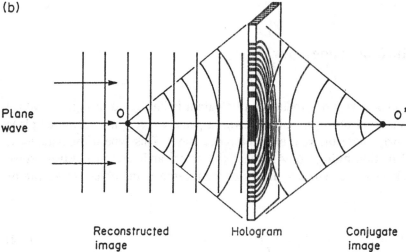

Fig. 2.1a,b. In-line holography of a point object: (a) Hologram formation, and (b) image reconstruction

The interference pattern given by (2.5) consists of concentric fringes and is called a **zone plate**.

If this hologram is recorded on film with a contrast of $\gamma = -2$ and is thereafter illuminated with a plane wave identical to the reference wave, it is possible to express the resultant transmitted amplitude $T(x,y)$ as

$$T(x,y) = e^{ikz}\left[1 + \left(\frac{f}{l}\right)^2 + i\frac{f}{l}\exp\left(\frac{ik(x^2+y^2)}{2l}\right)\right.$$
$$\left. - i\frac{f}{l}\exp\left(\frac{-ik(x^2+y^2)}{2l}\right)\right]. \tag{2.6}$$

The first and second terms represent the transmitted plane waves. The third term gives the original wave scattered from the point object, that is, a spherical wave diverging from the point O (Fig.2.1b). The fourth term represents a spherical wave converging to the point O′, located at the position mirror-symmetric to the point O with respect to the hologram plane. This fourth term describes the **conjugate image**.

In short, the hologram of a point object, of which the hologram is a zone plate, plays dual roles of concave and convex lenses with the same focal length l. Illuminating this hologram with a plane wave therefore produces both divergent and convergent spherical waves.

We can now discuss the resolution of the reconstructed image. An in-line hologram acts as a lens, and a point image is formed by the interference of all the waves diffracted from the zone-plate fringes. The image resolution d is therefore determined essentially by the diameter D of the zone plate, which corresponds to the diameter of the lens aperture, i.e.,

$$d = 1.6\frac{\lambda}{D}l. \tag{2.7}$$

This value of d is equal to the shortest fringe spacing at the outermost fringe of the zone plate.

Generally, it is not possible to observe only the reconstructed image: the defocused pattern of the conjugate image is inevitably superimposed onto the reconstructed image. This results from the fact that both images lie on the same axis. The problem of separating the twin images was a persistent obstacle to holography. A solution was found, however, by introducing a new method called **off-axis holography** [2.2]. As will be discussed in Sect.2.2, a reference wave is tilted with respect to an object wave.

Although off-axis holography could solve the problem of a conjugate image, further efforts were made to reconstruct in-line holographic images free from distortions. The most effective of these methods is the **Fraunhofer in-line holography**. Here, in-line holograms are formed in the Fraunhofer-diffraction plane of an object, that is, under the condition

$$a^2 \ll \lambda l, \tag{2.8}$$

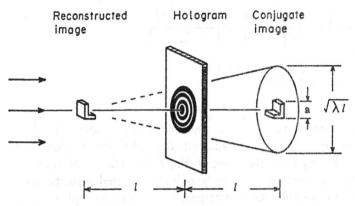

Fig. 2.2. Fraunhofer in-line holography. When a hologram is formed in the Fraunhofer diffraction region of an object (2.8), the reconstructed image is not disturbed by the conjugate image. The dimensions are reduced to those in the hologram-formation process

where a is the size of the object. Under this condition, reconstructed twin images have no influence on each other. If one of the twin images is observed, the other image is completely blurred in this plane (Fig. 2.2). Simple calculations indicate that near the region of the reconstructed image, the intensity of the blurred image is a small constant value [2.3].

2.2 Off-Axis Holography

Off-axis holograms are formed between object and reference waves propagating in different directions. If a reference wave is assumed to be a tilted plane wave represented by $\phi_r = \exp[ik(z-\alpha y)]$, then the intensity I at the hologram plane is given by

$$I(x,y) = 1 + |\phi_0|^2 + \phi_0 e^{-ik(z-\alpha y)} + \phi_0^* e^{ik(z-\alpha y)} \ . \qquad (2.9)$$

For a point object, that is, when $\phi_0 = \mathrm{i} f r^{-1} e^{ikr}$ (Fig. 2.3a), the intensity distribution is

$$I(x,y) \approx 1 + \left(\frac{f}{l}\right)^2 - \frac{2f}{l}\sin\left(\frac{k(x^2+y^2)}{2l} + k\alpha y\right) \ . \qquad (2.10)$$

6

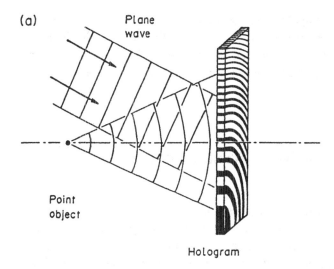

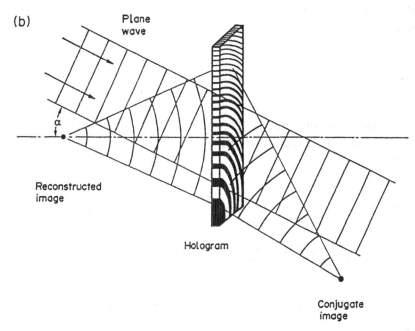

Fig. 2.3a,b. Off-axis holography of a point object: (**a**) Hologram formation, and (**b**) image reconstruction

Whereas the *in-line hologram* of a point object is a zone plate consisting of concentric circular fringes, this *off-axis hologram* can be considered a zone plate at some distance from the center of the zone plate, and to be an almost sinusoidal grating that is slightly modulated by an object.

If a tilted plane wave $\exp[ik(z-\alpha y)]$ illuminates the off-axis hologram, represented by (2.9), the original object ϕ_0 is reconstructed. The propagation direction of the conjugate image is tilted by the angle 2α with respect to the direction of the reconstructed image (Fig.2.3b). As a result, the twin images are spatially separated and can be observed independently without distortion. The imaging principle of off-axis holography is the same as that for in-line holography.

2.3 Holography Using Two Different Kinds of Waves

Up to now, we have considered only cases in which the same kind of waves are used to form the hologram and reconstruct the image. However, holography is not restricted to this condition, but is valid for arbitrary waves. To take a simple example, a zone plate (2.5) is magnified m times and then illuminated by a wave having a different wave number k'. The intensity of the enlarged hologram is calculated to be

$$I(x,y) = 1 + \left(\frac{f}{l}\right)^2 - \frac{2f}{l}\sin\left(\frac{k[(x/m)^2 + (y/m)^2]}{2l}\right). \tag{2.11}$$

When the hologram is illuminated with the wave having the wave number k', the transmitted amplitude on the plane at the distance l' from the hologram is given by

$$T(x,y) = e^{ik'l'}\left[1 + \left(\frac{f}{l}\right)^2 + i\frac{f}{l}\exp\left(\frac{ik[(x/m)^2 + (y/m)^2]}{2l}\right)\right.$$
$$\left. - i\frac{f}{l}\exp\left(\frac{-ik[(x/m)^2 + (y/m)^2]}{2l}\right)\right]. \tag{2.12}$$

This equation has the same form as (2.6) if l' is selected to satisfy the relation

$$l' = m^2 \frac{\lambda}{\lambda'} l. \tag{2.13}$$

This means that when a wave other than that used to form the hologram illuminates the hologram, the original wave front is reconstructed in a slightly modified form, with a longitudinal magnification of $m^2\lambda/\lambda'$ and a lateral magnification of m. The lateral magnification m is evident if one considers the reconstruction of two point objects: the distance between the two zone-plate centers in the hologram is equal to the distance between the two reconstructed point images. The longitudinal and lateral magnifications are the same when

$$m = \lambda'/\lambda. \tag{2.14}$$

Then the reconstructed wave front is simply scaled up by the wavelength ratio λ'/λ.

Up to this point, simple discussions have demonstrated that holography is possible between two arbitrary waves even though their wavelengths and wave properties are different. This is possible because holography uses only the most fundamental features, *interference* and *diffraction*, which are common to all kinds of waves.

3. Electron Optics

This chapter briefly introduces electron optics, emphasizing those aspects that are important for holography.

3.1 Electron Microscopy

The idea of electron holography was conceived during the development of electron microscopes. Furthermore, electron holograms are usually formed in electron microscopes. We should therefore first consider the basic concept of a transmission electron microscope [3.1].

3.1.1 Ray Diagram

An electron microscope consists of an electron gun, an illuminating lens system, a specimen chamber, a magnifying lens system, and a recording system (Fig. 3.1). The electron beam emitted from the electron gun is collimated by condenser lenses and then illuminates the specimen. The accelerating voltage ranges from 100 kV to 3.5 MV, and the illumination angle is typically on the order of 10^{-3} rad.

The electron beam transmitted through the specimen forms a magnified specimen image first through the objective lens and then through magnifying lenses: the total magnification may reach a factor of 1,000,000. The magnified image is directly observed on a fluorescent screen, or a TV camera, or it is recorded on film. The optical system described thus far is very similar to that of an optical microscope, but each component is quite different from its optical analogue.

3.1.2 Electron Guns

The large chromatic aberrations of electron lenses require that an electron beam must be as monochromatic as possible. In addition, the brightness of

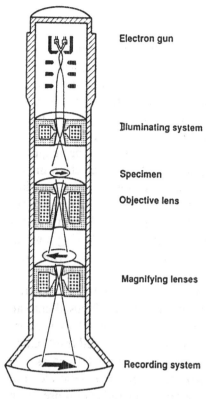

Fig. 3.1. Ray diagram of an electron microscope

Electron gun

Illuminating system

Specimen

Objective lens

Magnifying lenses

Recording system

an electron beam must be high enough to produce an image that is still visible after being magnified as much as 1,000,000 times its original size. The characteristics of various electron guns will be compared in Sect. 3.3 in relation to the coherence properties of electron beams. Various types of electron guns have been developed and can broadly be classified into two types: **thermal-emission** or **field-emission**. Typical thermal-emission guns consist of a hair-pin tungsten wire, a tungsten needle attached to the hairpin wire (pointed filament) [3.2], or a LaB_6 needle [3.3] from the pointed ends of which thermal electrons are extracted by an applied electric field.

In a field-emission gun [3.4], electrons are emitted from an unheated sharp tip only through the application of a high electric field. The tip radius is about 100 nm, whereas the radius of a thermal pointed filament is about 1 μm. To obtain stable emission, the pressure of the gun chamber must be 10^{-8} Pa or less. Stable emission can be obtained even at a pressure of 10^{-6} Pa, when the tip is heated to 1800 K, however, the monochromaticity of the electron beam degrades.

3.1.3 Electron Lenses

Electron lenses are quite different from optical lenses: an optical lens is usually made of glass, whereas an electron lens uses electromagnetic fields. This is because electrons cannot pass through thick glass due to the strong interaction of electrons with matter. In addition, the refractive index of the substance (ratio of the electron velocities in the vacuum and the substance) is very near to that of vacuum such that the ratio is typically 1.00001.

Electron lenses are classified into two types: electrostatic or magnetic. Both types are widely used for different purposes. Modern electron microscopes often use magnetic lenses, mainly because this kind of lens has smaller aberrations [3.5]. The key component of a magnetic lens is a circular coil through which magnetic flux flows (Fig. 3.2). The coil is surrounded by a toroidal iron yoke that has a gap inside. Magnetic flux flowing through the yoke leaks at the gap. The leakage magnetic field is axially symmetric and a function of the electron lens.

Following the paths described by paraxial ray equations, all electron rays in the magnetic field starting from an object point are focused onto an image point. A slight difference between electron and optical lenses is that electron rays are not confined to a meridional plane but instead travel along a spiral. It is interesting to note that electron wave fronts, defined as perpendicular to electron trajectories, are also spirals in a magnetic electron lens. The aberrations of a magnetic electron lens are similar to those of an optical lens, except for the additional aberrations associated with the spiral motion.

The paraxial-ray equation demonstrates that no concave electron lenses exist. This is the reason why the spherical and chromatic aberrations in an electron microscope cannot be compensated by combining convex and concave lenses. This lack of compensation places some limitations on electron

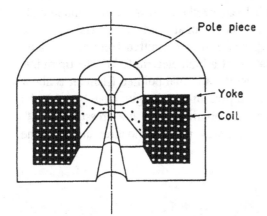

Fig. 3.2. Magnetic electron lens

microscopes. Their image resolution is restricted by the spherical and chromatic aberrations of the objective lens, to be 100 to 300 pm which is much larger than the wavelength of 4 pm for a 100 keV electron beam.

Throughout this book I have used the unit picometer (pm) rather than Ångstrom (Å) though the latter unit seems more familiar in electron microscopy (1 Å = 100 pm).

3.2 Interference Electron Microscope

An interference electron microscope is an electron microscope that is equipped with an **electron biprism**. This microscope can measure such quantities as inner and contact potentials, the microscopic distribution of electric and magnetic fields, specimen thickness and trapped magnetic flux. The most important component in this kind of microscope is the electron biprism, that had been developed into the only practical interferometer by *Möllenstedt* and *Düker* [3.6].

The biprism is composed of a fine filament bridged in the center, and two plate-shaped electrodes that are both at ground potential (Fig. 3.3). The diameter of the filament has to be small enough to not disturb the narrow coherent region of the incident electron beam. Filaments should be less than 1 µm in diameter, and several methods have been developed for producing such fine filaments. In one method, a filament is picked up by extending a

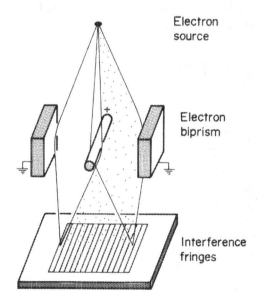

Fig. 3.3. Electron biprism

burnt rod of quartz, and a thin thread is caught by a tiny fork with two prong wires. The surface of the filament is coated by evaporating a layer of gold onto it. A similar filament-production method has been described in [3.7], and a method using a Wollaston wire as a biprism filament has been developed [3.8].

When a positive electric potential V_b ranging from 10 to 100 V is applied to the filament, two electron beams on opposite sides of the filament are attracted towards the filament and overlap on the lower plane. If the two beams are coherent, an interference pattern is produced there. This construction acts as a biprism and can be verified as follows. The electric potential $V(\rho)$ around the filament is approximately given by

$$V(\rho) = V_b \frac{\log(\rho/\rho_2)}{\log(\rho_1/\rho_2)}, \qquad (3.1)$$

where ρ_1 is the radius of the filament, and ρ_2 the distance from the filament center to the grounded electrode. If parallel electrons are incident on both sides of the filament, they are slightly deflected by the electric field toward the filament. When the deflection angle δ is small, it can be calculated as

$$\delta = \frac{\pi V_b}{2V \log(\rho_1/\rho_2)}, \qquad (3.2)$$

where V is the accelerating voltage of the electron beam. This equation shows that the deflection angle is constant, irrespective of the incident position of the electron beam, and that it is proportional to the voltage applied to the filament. This indicates that the electron biprism is a precise electron

Fig. 3.4. Electron interferogram of magnesium oxide particles

analogue of an optical biprism. If an object to be investigated is placed in one of the two interfering beams, an **interferogram** is obtained. An example of an interferogram with a specimen of magnesium oxide particles is presented in Fig. 3.4.

3.3 Coherence Properties of Electron Beams

The quality of an electron beam plays an important role in electron-holography experiments. In fact, the advent of the field-emission electron beam made it possible to directly observe an electron interference pattern on a fluorescent screen. The total number of interference fringes that can be recorded on film has been increased by an order of magnitude. However, the coherence of an electron beam is still not comparable to the coherence of a laser beam. For example, the average distance between coherent electrons is extremely large, say, 100 m even in the case of a coherent field-emission electron beam. An electron emitted from the electron gun travels about 1 m in the microscope, and is then detected on film. The next elec-

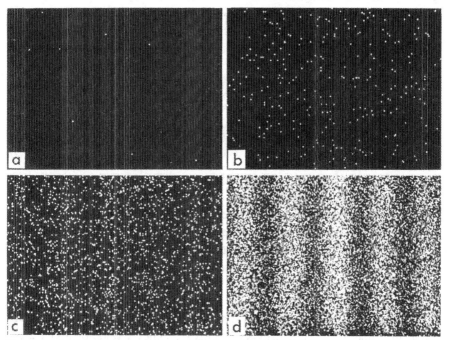

Fig. 3.5a-d. Single-electron build-up of biprism interference pattern: (**a**) 8 electrons, (**b**) 270 electrons, (**c**) 2000 electrons, and (**d**) 60,000 electrons

tron is produced after an interval being 100 times longer than the lifetime of the first electron. Figure 3.5 depicts the single-electron buildup of an electron interference pattern [3.9].

The coherence properties of an electron beam have been treated exactly by *Hawkes* [3.10], but can be separately considered as **temporal coherence** (longitudinal coherence) and **spatial coherence** (transverse coherence). Temporal coherence depends on the energy spread of the beam, and refers to the coherence in the propagation direction of a wave packet. Spatial coherence is influenced by the source size or the illumination angle, and refers to the coherence in the plane perpendicular to the propagation direction. The factors actually determining the two kinds of beam coherence, and values of the coherence lengths will be discussed in Sects. 3.3.1-2.

3.3.1 Temporal Coherence

The **temporal-coherence length** is the length of a wave packet in the direction of propagation. When an electron beam is coherently split into two beams and they then overlap after traveling along different paths, interference fringes can be observed if the difference between the path lengths is less than this temporal coherence length. The length l_t of a wave packet is given by the uncertainty $\Delta\lambda$ in the wavelength λ, or by the uncertainy $e\Delta V$ in electron energy eV, as expressed by

$$l_t = \frac{\lambda^2}{\Delta\lambda} = \frac{2V}{\Delta V}\lambda. \tag{3.3}$$

The number n of wavelengths contained in a single wave packet is given by

$$n = \frac{\lambda}{\Delta\lambda} = \frac{2V}{\Delta V}. \tag{3.4}$$

The values of $\Delta\lambda$ or ΔV depend on the types of electron guns producing the beam. Thermal electrons have an energy spread of approximately 1 kT, k being the Boltzmann constant and T the absolute temperature. When T = 3000 K, the energy spread theoretically amounts to 0.3 eV. However, because of the *Boersch* effect [3.11], the actual energy spread of an electron beam accelerated to, say 100 keV, becomes larger than the theoretical value.

Temporal coherence lengths for various types of electron beams are listed in Table 3.1. The energy spread $e\Delta V$ reaches a value of about 2 eV for the hairpin-type thermal cathode conventionally used in electron micro-

Table 3.1. Temporal coherence of electron beams

	Structure	Source size [μm]	Energy spread e ΔV [eV]	Temporal coherence length l_t [μm]	Wave number in a packet n
Hairpin cathode		30	2	0.4	10^5
Pointed cathode		1	1	0.7	$2 \cdot 10^5$
Field-emission cathode		$5 \cdot 10^{-2}$	0.3	4	10^6

scopes. A thermal electron beam from a pointed cathode, developed by *Hibi* [3.2], has an energy spread of about 1 eV. The value of e ΔV decreases to about 0.3 eV for a field-emission electron beam, whose energy spread is given not by kT but by k'F (k' denoting a constant dependent on the work function, and F the electric field at the metal surface).

Special contrivances have been used to actually measure temporal coherence lengths. By employing a **Wien filter**, *Möllenstedt* and *Wohland* [3.12] measured the coherence length of a 2500 eV electron beam to be 1.12 μm. *Schmid* [3.13] later exhibited an electrostatic cylinder to measure a coherence length of 0.46 μm for a 35 keV electron beam. *Nicklaus* and *Hasselbach* [3.14] showed that the displacement of two partial wave packets could be restored by applying a Wien filter.

3.3.2 Spatial Coherence

The **spatial-coherence length** is the maximum distance between two points, in a plane perpendicular to the electron beam, where interference can occur. This length is the width of a wave packet and is determined by the illumination angle of an electron beam. A simple case is depicted in Fig. 3.6. Three kinds of parallel electron beams propagate in slightly different direction. The wave fronts perpendicular to the three directions are drawn so that they are in phase at point P0 in the observation plane. The wavefront displacement for the three beams becomes larger as the lateral distance from point P0 increases. The displacement of two beams, A and B or A and C, amounts to half a wavelength at point P1. Definite wave fronts can be observed within the region of radius ½P0P1, and the width l_s of this region can be roughly calculated as

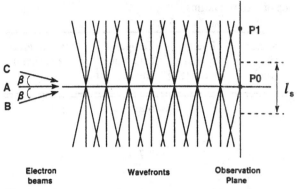

Fig. 3.6. Wave fronts of an electron beam with a nonzero illumination angle

$$l_s = \frac{\lambda}{2\beta}, \qquad (3.5)$$

where 2β is the illumination angle of the beam. An electron beam with an illumination angle 2β can thus be seen to have a spatial coherence length of approximately $\frac{1}{2}\lambda/\beta$.

It may seem that the coherence length can be made arbitrarily large by decreasing the illumination angle 2β, but in actuality this coherence length is restricted by the beam intensity required for observation, that is, by how large an intensity can be obtained within a small illumination angle. In electron optics, such characteristics are conventionally represented by the **brightness** B, which is defined as

$$B = \frac{i}{\pi \beta^2}. \qquad (3.6)$$

Table 3.2. Spatial coherence of electron beams

	Brightness B [A/(cm² · sr)] at 100 kV	Observable fringe number	Observed fringe number
Hairpin cathode	$5 \cdot 10^5$	150	100
Pointed cathode	$2 \cdot 10^6$	300	300
Field-emission cathode	$3 \cdot 10^8$	4000	3000

Here i is the current density of an electron beam, and $\pi\beta^2$ is a solid angle. It can be proven from Liouville's theorem that the value of B depends only on the kind of electron beam and the accelerating voltage, and that B is constant in any cross section of an electron beam. Spatial-coherence lengths of various electron beams are listed in Table 3.2. The brightness of an electron beam from a pointed filament is a few times that of a conventional hairpin-type gun, and a field-emission electron beam is nearly 1000 times brighter.

4. Historical Development of Electron Holography

The original approach that *Danis Gabor* took in developing electron holography was an in-line projection method [4.1] in which an object is illuminated with a divergent spherical wave from a point focus close to the object (Fig. 4.1). This kind of hologram is a highly magnified projected interference pattern between the object and the transmitted waves. No lenses are necessary, but this method requires a very small electron point focus. The diameter of the focus determines the resolution of the reconstructed image.

Because it was difficult in *Gabor*'s time to make a sufficiently fine probe, *Haine* and *Dyson* [4.2] devised a more practical transmission method so that holograms could be formed easily in electron microscopes. However, recently the projection method has been revived with a single-atom field-emitter [4.3, 4]. In this kind of experiment a low-energy, bright electron beam (a few hundred eV) field-emitted from a single-atom tip illuminates an object located near the tip to form a greatly enlarged interference pattern. No electron lenses are needed for forming a fine electron probe or for enlarging the interference pattern. This technique may open up applications different from the present technique with a high-energy electron beam, as described in this book.

Holograms formed by means of the in-line transmission method represent defocused electron micrographs photographed under coherent illumination. Such a micrograph was already possible to be obtained even in the early days of research in this field: **Fresnel fringes** in a largely defocused photograph of a specimen edge had been reported by *Boersch* [4.5].

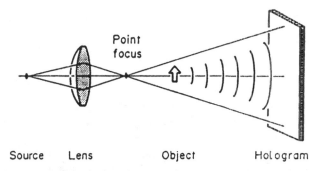

Fig. 4.1. In-line projection holography

Optical experiments demonstrating the feasibility of holography were reported by *Gabor* [4.1], *Rogers* [4.6], *Baez* [4.7], and *Kirkpatrick* and *El-Sum* [4.8]. The first experiment on "electron" holography was carried out by *Haine* and *Mulvey* [4.9], who formed a hologram as a defocused image of zinc-oxide crystals in which many Fresnel fringes from crystal edges could be observed. Since this hologram was not formed in a Fraunhofer-diffraction plane, the reconstructed image was disturbed by the conjugate image. Actually Fresnel fringes that form the conjugate image can be seen around the reconstructed image. Nevertheless, the reconstruction was verified, although not perfectly. Similar results were obtained by *Hibi* [4.10], who developed a **pointed cathode** for use as a coherent electron source, and reconstructed images from the Fresnel fringe-like holograms formed by utilizing this source.

4.1 In-Line Holography

Although the conjugate image problem was solved by using off-axis laser holography [4.11], *DeVelis* et al. [4.12] demonstrated that disturbance-free image reconstruction is possible even via **in-line holography** if holograms are formed in a Fraunhofer-diffraction plane of an object. An electron version of this experiment was carried out by *Tonomura* et al. [4.13] and clear images were reconstructed for the first time. In this experiment a 100-kV electron microscope equipped with a pointed cathode was employed (λ = 3.7 pm). The electron-optical system for hologram formation is illustrated in Fig. 4.2a. An electron beam from a pointed cathode was focused through the first condenser lens and was then collimated to illuminate a specimen through the second condenser lens. Small objects surrounded by clear spaces were selected as specimens so that the transmitted electron wave could be regarded as a coherent background. The diffraction pattern of a specimen under coherent illumination at a plane l = 2 mm distant from the specimen was magnified by 3000 to 6000 times by electron lenses, and recorded on film. Since the illumination angle 2β was made as small as 1×10^{-6} rad, the low current density necessitated an exposure time of up to ten minutes. The film was reversed with high contrast and used as a hologram.

Optical reconstruction was carried out with a HeNe laser (λ' = 632.8 nm). Just as in the hologram-formation process, light from the laser was focused and collimated through two optical lenses to illuminate an electron hologram (Fig. 4.3b). Two images were reconstructed at a distance of l' = $m^2 l \lambda / \lambda'$ from the hologram. When l = 2 mm, λ = 3.7 pm, λ' = 632.8 nm and m = 3000, the distance l' can be calculated to be about 10 cm.

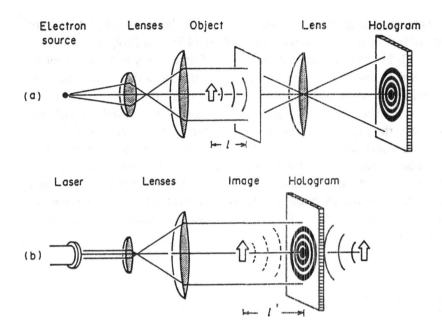

Fig. 4.2a,b. In-line transmission holography: (a) Hologram formation, and (b) image reconstruction

An example of the reconstruction is depicted in Fig. 4.3. The objects are fine particles of gold placed on a thin carbon film. The diameter a of the smaller particles is about 10 nm. The Fraunhofer condition is satisfied for these objects. In fact, the distance $l = 2$ mm from object to hologram is sufficiently large compared with $a^2/\lambda = 27$ μm to satisfy (2.8). Under this condition, holograms consist of concentric interference fringes (zone plate) irrespective of the shape of an object, as evidenced by Fig. 4.3a.

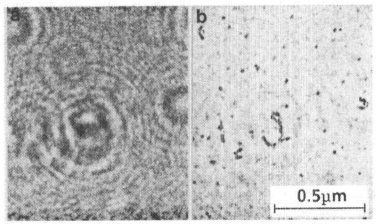

Fig. 4.3a,b. Fine gold particles: (a) Electron hologram, and (b) reconstructed image

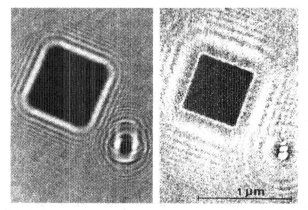

Fig. 4.4a,b. Magnesium oxide particle: (a) Electron hologram, and (b) reconstructed image

The reconstructed image in Fig. 4.3b demonstrates that this Fraunhofer condition negates the image disturbance due to the conjugate image. This is the result of the conjugate image being spread all over a region whose diameter is $(2\lambda l)^{1/2}$, at the plane of the reconstrcuted image, which is much larger than the object size under the Fraunhofer condition. (It should be noted that the lengths here are reduced to those in the hologram-formation stage in an electron microscope).

Another example, where the specimen is a magnesium-oxide crystal, is depicted in Fig. 4.4. Since one side of the crystal is 600 nm in length, the Fraunhofer condition is not satisfied. This can be seen merely by looking at this hologram, since the interference fringes produced are Fresnel fringes and differ from the interference fringes of a zone plate. The reconstructed image (Fig. 4.4b) is therefore surrounded by Fresnel fringes from the conjugate image formed at the mirror-symmetric position with respect to the hologram plane. However, in this case the disturbance is produced only outside the particle image, and the particle edge is reconstructed undisturbed. Holes in the supporting collodion film, which have a negligible contrast in the electron micrograph, have clearly been reconstructed because of the higher contrast ($|\gamma| > 2$) of this hologram. The hole size satisfies the Fraunhofer condition.

The resolution d was a few nm in the image reconstructed in the above experiment. This value is determined by the diameter D of the zone plate recorded on a hologram, as was shown by (2.7). This diameter is determined by the spatial coherence length l_s of the illuminating electron beam, whose length is given by $\lambda/2\beta$ in terms of the illumination angle 2β. Consequently the resolution d can be represented by

$$d = 3.2\beta l .\qquad(4.1)$$

This relationship indicates that high resolution can be achieved when β and l are small. In this experiment, d is calculated to be 3 nm by using $2\beta = 10^{-6}$ rad and $l = 2$ mm, which is consistent with the experimental results.

Hanszen [4.14] theoretically investigated the contrast transfer function of holographic imaging and proposed a method for improving the resolution of electron microscopes. If the image of a phase object is reconstructed from the bleached weak-phase hologram in an optical system whose aberrations and defocusing are the same as those in the hologram-formation step, the transfer function of in-line holography becomes the square of that for an electron lens. This removes the contrast-inversion effect at some spatial frequency regions, that always accompanies a highly magnified electron micrograph because of the spherical aberration in the electron lens. The remaining problem of frequency gaps can be removed by superposing several

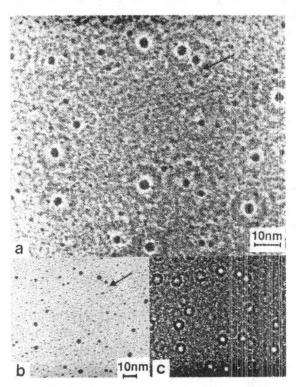

Fig. 4.5a-c. High-resolution electron holography [4.15]: (**a**) Reconstructed image of gold particles, (**b**) electron micrograph, and (**c**) electron hologram

images with different defocusings, the transfer intervals of which replenish each other.

The resolution of reconstructed images was improved by *Munch* taking advantage of a field-emission electron beam [4.15]. The high brightness of this electron beam allowed the illumination angle 2β to be reduced. In addition, finer particles were selected as samples, thereby allowing a smaller sample-hologram distance l. A resolution of 1 nm was obtained for $l = 5$ μm (Fig.4.5). *Munch* concluded that higher-resolution imaging was difficult to achieve with in-line holography, since the contrast of the zone plate is extremely low because of the weak intensity of the wave scattered from a small object like an atom. *Bonnet* et al. [4.16] also used a field-emission electron beam to obtain an image resolution of 700 pm.

The in-line holography described up to this point is constrained by several limitations, namely that the objects be small and surrounded by clear spaces, etc. However, the coherence conditions for an illuminating electron beam are much less stringent than those for off-axis holography. In-line holography is therefore suited to specimens such as isolated fine particles, although there are still problems with regard to the supporting film through which an electron beam should transmit as a coherent background wave.

4.2 Off-Axis Holography

Off-axis electron holography was first carried out by *Möllenstedt* and *Wahl* [4.17]. They used a slit-shaped electron source, aligning the biprism direction so that as many biprism interference fringes as possible could be recorded on film. An image of a thin tungsten filament was optically reconstructed (Fig.4.6). The reconstructed image is resolved only in a direction perpendicular to the filament. *Tonomura* [4.18] subsequently reconstructed an image using a single-crystalline thin film as an amplitude-division beam splitter.

Weingärtner et al. [4.19] predicted that an image resolution of as high as 40 pm could be attained by using off-axis "image" holography. They concluded that the image resolution was less affected by the illumination angle of the electron beam in this type of holography than in any other type of off-axis holography. The spherical aberration of an electron lens is compensated for in the reconstruction stage by that of an optical concave lens to improve the resolution.

The use of off-axis Fresnel holography to improve the quality of the reconstructed image was reported by *Tomita* et al. (Fig.4.7) [4.20, 21] and by *Saxon* [4.22, 23]. The former used a pointed cathode and the latter a field-

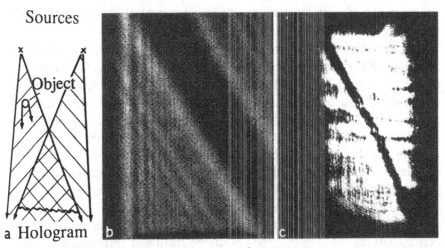

Fig. 4.6a-c. Off-axis electron holography [4.17]: (a) Schematic diagram of hologram formation, (b) electron hologram, and (c) reconstructed image

emission electron gun. *Saxon* further demonstrated how coma aberrations in an electron lens can be corrected during the optical reconstruction stage.

The image quality was greatly improved by *Tonomura* et al. [4.24], who employed a field-emission electron beam that could produce 3000 biprism fringes instead of the 300 previously possible. The reason for the improvement lies in the fact that the number of carrier fringes in a hologram corresponds to the number of picture elements. This electron beam could be

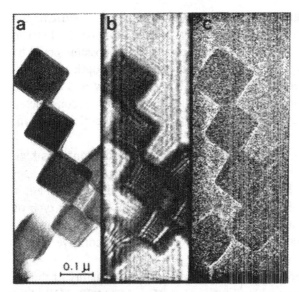

Fig. 4.7a-c. Off-axis electron holography [4.20]: (a) Electron micrograph of magnesium oxide particles, (b) electron hologram, and (c) reconstructed image

used to reconstruct lattice fringes of gold {111} planes (spacing: 240pm) together with the half-spacing fringes [4.25]. Even Bragg-reflected beams could be reconstructed optically, and the resolution of a reconstructed image became comparable to that of an electron microscopic image. The spacing between carrier fringes in this experiment ranges from 50 to 200 pm. The narrow spacing of the carrier fringes is indispensable for high-resolution holography, since the resolution of the reconstructed image is about three times this spacing. *Lichte* [4.26] was able to reconstruct carbon-black lattice fringes with a spacing of 340 pm when the spacing of carrier fringes was 80 pm.

More recent developments in high-resolution off-axis holography are described in detail in Chap. 8, and only outlined here: a structure image of Nb_2O_5 was also reconstructed by *Lichte* [4.27]. *Völkl* and *Lichte* [4.28] formed structure-image holograms of cerium dioxide crystals, and these holograms had a carrier fringe spacing of 32 pm. By using a 350 kV field-emission electron microscope [4.29], *Kawasaki* et al. [4.30] reconstructed a structure image from 150 diffracted waves and demonstrated the formation of a high-resolution hologram with a carrier fringe spacing as narrow as 9 pm. *Tanji* et al. [4.31] investigated the surface structure of MgO in detail by observing the holographically obtained phase image in which the effect of Fresnel fringes accompanied in the profile-mode observation was reduced. Thus, the resolution of an electron holography has become comparable to that of electron microscopy. In the near future, higher resolutions may be obtained by holographically compensating for the aberrations in electron lenses.

Off-axis holography without an electron biprism was investigated by *Matteucci* et al. [4.32]. A single-crystalline film was used as a wave-front-division beam splitter. An electron beam incident on the film excited the Bragg reflection to split into two (transmitted and reflected) beams. When

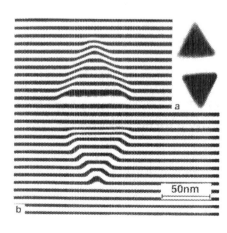

Fig. 4.8a,b. Interence micrograph of fine gold particles on a molybdenite thin film reconstructed from an incoherent hologram. The thickness distribution can be measured quantitatively, since one fringe shift corresponds to a thickness change of 33 nm for a mean inner potential of 40 V at a 200 kV electron beam

an object was placed in the path of one of the two beams, a lattice image modulated by the object was formed at the image plane of an electron lens. This image can thus be regarded as an off-axis Fresnel hologram. The coherence conditions for an illuminating electron beam are less stringent than those when an electron biprism is utilized. Consequently, holograms can be formed using even an electron beam from a pointed cathode.

This incoherent holography was employed by *Ru* et al. [4.33, 34] to obtain interference micrographs. An example is depicted in Fig. 4.8, where fine gold particles epitaxially grown on a molybdenite film were reconstructed from the incoherent hologram.

Another interesting method is Fourier-transform holography, as was demonstrated by *Lauer* [4.35]. A weak scattering foil was placed at the normal specimen position of an electron microscope, and an object was placed at the back focal position of the objective lens (Fig. 4.9). The object was restricted to one half the focal plane; the other half was screened off by a mask. The resolution of the reconstructed images was 3 nm.

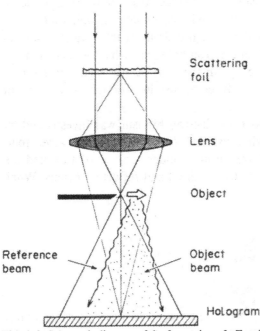

Fig. 4.9. Schematic diagram of the formation of a Fourier-transform hologram

5. Electron Holography

This chapter introduces typical examples for the electron holography techniques that have hitherto been developed.

5.1 Electron-Hologram Formation

Electron holography consists of two steps: Electron-hologram formation and image reconstruction. In the first step, off-axis electron holograms are formed in an electron holography microscope which is essentially a transmission electron microscope [5.1] with a field-emission electron gun and an electron biprism [5.2].

5.1.1 Ray Diagram

A typical method for forming an off-axis image hologram is detailed in Fig. 5.1. An electron beam emitted from a field-emission tip is accelerated and then collimated to illuminate an object through a condenser lens system. The collimated illumination is necessary to obtain an illumination angle small enough for the spatial coherence length (3.5) to cover both the object field and the reference beam. For example, if we use a 200 keV electron beam (wavelength: 2.5pm) to observe an object region of 5 μm across, then the spatial-coherence length must be longer than 10 μm, and the illumination angle must be smaller than $2.5 \cdot 10^{-7}$ rad.

In this technique, an object is located in one half of the object plane, and is illuminated with a collimated electron beam. The image of the object is magnified ($\times m_1$) through an objective lens. When a positive potential is applied to the central filament of the electron biprism below the objective lens, the object image and the reference beam overlap each other to form an interference pattern. Since the interference-fringe spacing is as small as a few tens of nanometers, before the pattern is recorded on film, it is enlarged ($\times m_2$) by magnifying lenses.

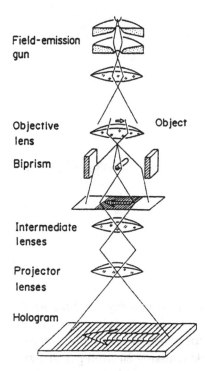

Fig. 5.1. Ray diagram of the formation of an electron hologram

The specimen magnification is given by $m_1 \times m_2$, and the magnification of the biprism interference pattern is given by m_2. When holograms are recorded on an electron-microscopy film, the spacing between fringes in a hologram usually ranges from 50 μm to 200 μm. Wide spacing is especially needed when the phase distribution is to be precisely reconstructed by using the phase-amplification technique, since a small phase shift is recorded as a slight deviation from equally spaced parallel fringes. The reduced fringe spacing in an object plane is given by d_b/m_1 (d_b is the biprism fringe spacing). This value is important in that the resolution of reconstructed images is approximately $3d_b/m_1$.

The values of m_1 and m_2 have to be adjusted to form holograms under the best conditions. However, we have to remember that the fringe spacing d_b is determined not only by m_1 but also by the potential applied to the biprism, which mainly determines the width of the biprism interference pattern. If desirable conditions cannot be found, the electron-optical system should be altered, for example, by moving the electron biprism to below the first intermediate lens.

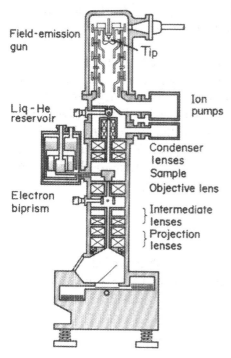

Fig. 5.2. Cross-sectional diagram of a holography electron microscope

5.1.2 Experimental Apparatus

The 200-kV holography electron microscope shown in Fig. 5.2 is essentially a Hitachi HF-2000 field-emission transmission electron microscope which has slightly been modified so that an electron biprism can be installed below the intermediate lens as well as below the objective lens.

a) Electron Gun and Illumination System

Electrons are emitted from a ⟨310⟩-oriented tungsten tip at room temperature simply by applying an electric potential of 3 to 6 kV between the tip and the first anode located just below the tip. The ambient pressure around the tip is kept at about 10^{-8} Pa so that a stable total emission current of up to 100 μA can be supplied continuously for several hours. Only a small fraction (1% or less) of the electrons passes through the first anode aperture, after which they are accelerated by the following electrodes. An acceleration tube is indispensable for achieving a long-life tip in a high-voltage gun, since large high-voltage discharges often destroy pointed tips. A 200-kV gun (HF-2000) has seven stages, whereas a 350 kV gun has ten stages.

Electrons are subjected to a strong convergence from the lens action when passing through the first anode aperture. The electrode shape has been designed to minimize spherical and chromatic aberrations in the accelerating electrostatic lens. When the beam is focused through a condenser system onto the specimen, the diameter of the focused spot is as small as 20 nm. The effect of aberrations on the spot size cannot be neglected: the intrinsically high brightness of a field-emission electron beam is always degraded, to some extent, by the aberrations of the accelerating lens [5.3]. Brightness is also affected by the condenser lenses. To make a hologram, an electron beam is collimated by a condenser lens system to illuminate an object. An electron biprism is used to overlap the object beam and a reference beam.

b) Electron Interferometer

The electron biprism is usually located below the objective lens. Two kinds of biprism devices are shown in Fig. 5.3. A biprism consists of a fine cen-

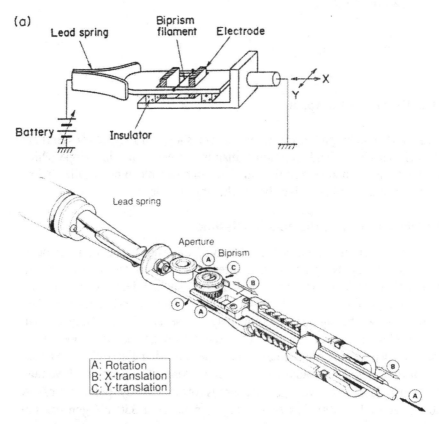

Fig. 5.3a,b. Electron biprism: (a) Translationally movable biprism, and (b) translationally and rotationally movable biprism

32

tral filament and two grounded parallel-plate electrodes, one on each side. A variable electric potential of −300 to +300 V is applied to the central filament through a lead spring from outside the microscope. The stability of this potential should be on the order of 10^{-6}/min, and dry cells are therefore often used to supply this potential. The central filament should be less than 1 μm in diameter so that the filament shadow does not obscure the extremely short spatial coherence length of the electron beam. When the filament is fabricated, as described in Sect. 3.2, its diameter can be made as small as 0.3 μm.

Like the objective aperture, the biprisms are movable in a plane perpendicular to the direction of the electron beam. However, with the biprism depicted in Fig. 5.3a the direction of the filament cannot be rotated. Usually there is no problem when an object can be rotated, but when an object cannot be rotated, for example, because it is on a low-temperature stage, it would be advantageous to be able to rotate the biprism. An example of such a biprism is illustrated in Fig. 5.3b. In this case the whole biprism stage can be rotated up to ±90°. The interference pattern between object and reference waves formed with the electron birprism is enlarged by intermediate and projector lenses, and it is recorded as a hologram on film or on videotape.

c) Recording System

To make a hologram of a weak phase object that is invisible even by electron microscopy, at least the existence of the object has to be detected as a displacement of an interference fringe. In an electron holography microscope to which a field-emission electron gun is attached, electron holograms can be directly observed on a fluorescent screen when there are fewer than 50 carrier fringes. The location of an object that produces only a small phase shift, however, can be more easily detected if the number of fringes is decreased from 50 to a few, since this makes the fringe spacing wider and the fringe intensity brighter.

Electron holograms are recorded on an electron-microscope film that has a resolution of 150 lines/mm. This resolution may seem low for holography, but it is a result of the brightness of an electron beam (current density per unit solid angle) being less than that of a laser beam. Because of the long exposure times required, film with a resolution of more than 1000 lines/mm has only limited applications. Holograms can be dynamically observed by attaching a television camera to an electron microscope, and they can be recorded on videotape.

5.2 Image Reconstruction

An optical image can be reconstructed in one of the two diffracted beams simply by illuminating an off-axis electron hologram with a collimated laser beam corresponding to the reference beam used for the hologram formation. This optical image is essentially the same as the original electron image: the amplitude and phase are the same for both images except for lateral and transverse magnifications. Once this image is photographed, only the intensity of an electron beam transmitted through the specimen is recorded. Therefore the resultant photograph is equivalent to the electron micrograph. However, since the optical image reproduces both amplitude and phase, more information is available than can be obtained from only an electron micrograph. For example, the phase distribution can be displayed as an interference micrograph by putting an optical interferometer in the optical reconstruction stage. The high-resolution image disturbed by spherical aberrations of the objective lens in an electron microscope can be restored to an aberration-free image. Some examples of the experimental methods are discussed in the following subsection.

The optical image-reconstruction process is considered to consist of three steps: Fourier transformation of a hologram, the selection of one sideband, and the Fourier transformation of the sideband. These steps can also be carried out numerically, though not instantly as in an optical reconstruction in time, depending on both the performance of the computer used for the image reconstruction and the number of pixels in the hologram.

5.2.1 Interference Microscopy

An **interference micrograph** can be obtained by overlapping a reconstructed image and a plane wave, as represented by Fig.5.4. A collimated light beam from a laser is split into two coherent beams by the beam splitter A. One of the two beams illuminates the electron hologram, forming twin images in the two diffracted beams. Lenses E and F form one of the images again on the observation plane: these two lenses form a confocal system so that a parallel beam incident onto the system leaves it as a parallel beam. The aperture at the back focal plane of the first lens E selects only the virtual image. The other beam from the splitter is sent directly to the observation plane as a comparison beam for forming the interference image.

A contour map of the phase distribution is obtained when two interfering beams are completely aligned in the same direction (Fig.5.5a). This map may be easy to interpret, especially if it is of a magnetic object, but lacks some wave-front information. For example, one cannot distinguish

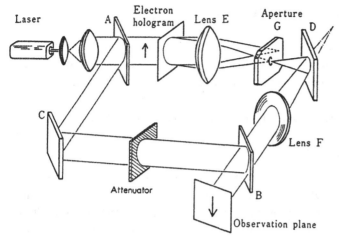

Fig. 5.4. Optical reconstruction system for interference microscopy

the contour map of a mountain from that of a valley. This means that the information obtained from a contour map alone cannot be used to determine whether a wave front is protruded or retarded. This information is available in an interferogram obtained by slightly tilting the direction of a comparison beam against that of the object wave (Fig. 5.5b). This is analogous to a mountain being more recognizable when viewed obliquely rather than from directly above.

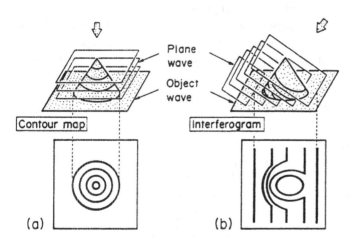

Fig. 5.5a,b. Interference micrographs: (a) Contour map, and (b) interferogram

5.2.2 Phase-Amplified Interference Microscopy

In conventional interference microscopy, the precision of the phase measurement is $2\pi/4$. However, electron phase shifts produced by an object are not always large enough to be measured. This is particularly true if an extremely small change in thickness or a small magnetic flux is to be observed. In these cases, we use a technique peculiar to holography.

a) Optical Method

The present technique making full use of the conjugate image was devised by *Matsumoto* and *Takahashi* for an optical field [5.4], and was first applied to electron holography by *Endo* et al. [5.5, 5]. The precision in phase measurement was demonstrated to be 1/50 of the electron wavelength using a mono-atomic step in a thin film as an object [5.7].

The principle underlying phase-amplified interference microscopy is illustrated in Fig. 5.6. A conventional interference micrograph is formed by the interference between a reconstructed image and a plane wave. The phase difference between the reconstructed wave front and the plane wave produces the interference micrograph (Fig. 5.6a). The new method takes advantage of a conjugate image instead of a plane wave (Fig. 5.6b). The phase difference exactly doubles, because twin images have amplitudes that are the complex conjugate of each other's amplitude: that is, their phase values are reversed in sign. As a result, contour fringes appear at every phase change of π as if the phase was amplified twofold.

The problem is how to superpose the twin images. An optical system to achieve this is not difficult to design, and an example of one is presented in Fig. 5.7. A Mach-Zehnder interferometer splits a collimated light beam

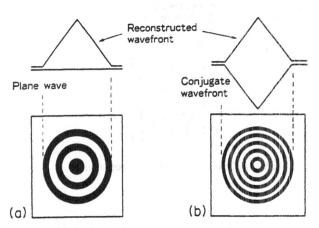

Fig. 5.6a,b. Principle behind phase-amplified interference microscopy: (**a**) Contour map, and (**b**) twice-amplified contour map

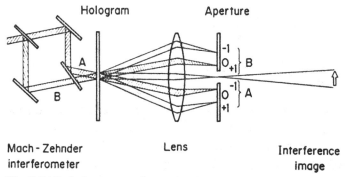

Fig. 5.7. Optical reconstruction system for phase-amplified interference microscopy

from a laser into two coherent beams traveling in different directions. These two beams illuminate an electron hologram and produce two sets of a reconstructed image and its conjugate. The reconstructed image of one beam and the conjugate image of the other beam are made to coincide by using the Mach-Zehnder interferometer to adjust the incident angles to the hologram. The phase distributions of these twin images are equal in absolute value but opposite in sign. This forms an interference image with a twice-amplified phase distribution.

More than twofold phase amplifications can be carried out by using higher-order diffracted beams produced from a nonlinear hologram. Although only $\pm 1^{st}$-order diffracted beams are produced from a linear hologram with the contrast $\gamma = -2$, higher-order beams can be obtained from a nonlinear hologram ($\gamma \neq -2$). However, the intensities of the higher-order beams are generally weak. In that case, holograms are often reversed with high contrast and/or bleached out so that higher-order diffracted beams may become more intense relative to the transmitted beam or the hologram transmittance may become higher.

When a reconstructed image of a Nth-order diffracted beam and a conjugate image of a $-N^{th}$-order diffracted beam are superposed, the amplification rate becomes 2N because an image reconstructed with a Nth-order diffracted beam has its phase distribution amplified N times. The reason for the phase amplification in higher-order diffracted beams can be explained as follows. Suppose that the intensity of an off-axis hologram is not given by (2.10), as in the case of a point object, but in general by

$$I(x,y) = 1 + \epsilon \sin[\psi(x,y) + k\alpha y], \qquad (5.1)$$

where $\psi(x,y)$ expresses the phase distribution of an object wave. If ϵ is less than unity, and the hologram is recorded with the contrast γ, then the transmittance t of the hologram, given by (2.2), can be expanded in the power series

$$t(x,y) = 1 - \frac{\gamma}{2}\epsilon \sin Q + \frac{\gamma(\gamma+2)}{8}\epsilon^2 \sin^2 Q$$
$$- \frac{\gamma(\gamma+2)(\gamma+4)}{48}\epsilon^3 \sin^3 Q + \dots, \qquad (5.2)$$

where $Q = \psi(x,y) + k\alpha y$. If $\gamma = -2$, then all the terms except for the first and second terms vanish. That means only $\pm 1^{st}$-order diffracted beams expressed by $\exp(\pm iQ)$ are produced when the hologram is illuminated with a plane wave. However, when $|\gamma| \gg 2$, the higher-order terms cannot be neglected. The term $\sin^N Q$ contains the factors $\exp(\pm iNQ)$, which correspond to $\pm N^{th}$-order diffracted beams. In the Nth-order beam, therefore, the phase distribution is amplified N times, as is the inclination angle α. Thus the phase distribution of a reconstructed image formed with the Nth-order diffracted beam is N times as large as that of the original image.

The measurement precision and the phase amplification rate are not directly related. The precision with which an electron phase distribution is measured is essentially determined by the precision of the recorded hologram. Even slight phase shifts have to be recorded as slight deviations from regular carrier fringes in this hologram. The amplification technique can only help to visualize small phase shifts. For high amplification a high-contrast hologram therefore has to be formed. For example, in order to do so, the spacing of the carrier fringes should be 100 μm or wider when the hologram is recorded on an electron-microscopy film.

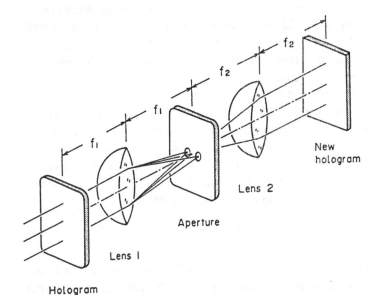

Fig. 5.8. Optical spatial filtering system for forming a phase-amplified hologram

Another way to obtain high amplification in interference microscopy is by repeating the phase-doubling technique by means of twin images. This reforms a twice phase-amplified interferogram with many carrier fringes (hologram), except for the last stage of repetition, rather than a phase-amplified contour map. For this purpose the simple yet high-precision optical system delineated in Fig. 5.8 is used instead of the Mach-Zehnder type interferometer. In this system a collimated laser beam illuminates a hologram, and $\pm 1^{st}$-order diffracted beams are selected by the aperture located at the back focal plane of lens 1. After this a twofold phase-amplified hologram is formed. If this procedure is repeated n times the resultant amplification rate is 2^n.

b) Numerical Method

An optical image can be reconstructed instantly from a hologram, but the preparation of the hologram is nonetheless time-consuming, especially with phase amplification, because the wet processes of the film development is slow [5.8]. An image can be reconstructed numerically from a hologram by carrying out a Fourier transformation twice [5.9, 10]: first, the Fourier transform of the hologram is obtained. This is equivalent to the optical diffraction pattern produced by shining a light beam onto the hologram plate. Thereafter, the Fourier transform of one side band of the optical diffraction pattern produces an image. The phase and amplitude of the image are independently obtained, and the phase distribution displayed either as a linear phase map (**contour map**) or as a sine-modulated interference micrograph (**interferogram**). Since the phase distribution is quantitively obtained, an n-times phase-amplified interference micrograph can be obtained simply by displaying the phase distribution in $2\pi/n$ units as a contour map.

c) Phase-Shifting Method in Optical Reconstruction

In addition to the above Fourier-transform method, a **phase-shifting** (or **fringe-scanning**) **method** [5.11] has also been proven effective for electron holography [5.12, 13]. These two methods can be used effectively for each specific purpose: the Fourier-transform method is more appropriate for real-time observations because an image can be reconstructed from one hologram. Although the phase-shifting method requires multiple holograms, which are formed under conditions of different initial phases of a reference wave, it can be used to measure the phase distribution more precisely.

An example of a phase-shifting reconstruction system [5.13] is shown in Fig. 5.9. A Twyman-Green interferometer splits a laser beam into two coherent beams traveling in different directions. Only the $\pm N^{th}$-order diffracted beams pass through the aperture to form a 2N-times phase amplified interference micrograph on a television camera. When the PieZoelectric

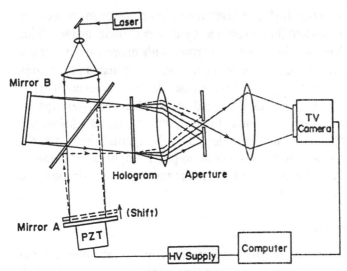

Fig. 5.9. Reconstruction system for phase-amplified interference microscopy using both optical and digital methods. Two coherent laser beams traveling in slightly different directions illuminate an electron hologram to form a phase-amplified interference micrograph monitored by a TV camera. Four different micrographs are obtained when mirror A moves in steps. The relative phase value at each pixel of the micrograph is calculated from the four intensity values and displayed as a high-precision phase-amplified contour map

Transducer (PZT) translates a mirror A in steps, the fringes in the interference micrograph are shifted. For example, interference micrographs at four different mirror positions at intervals of $\lambda/8$ (λ being the laser wavelength) are synchronously stored in an image processing computer by using a television camera. The phase value of each pixel in the image is calculated from the brightness values of the same pixel in the four micrographs. The original electron wave front is thus reconstructed numerically and is displayed as an arbitrarily amplified interference micrograph. The precision in phase measurements reached $\lambda/100$ [5.13].

d) Phase-Shifting Method in Electron Microscopy

All the methods introduced so far have been used for reconstructing phase-amplified interference micrographs from a single electron hologram. However, the phase-shifting method can directly be applied to an electron microscope. This technique provides a higher precision in phase measurements than does the conventional technique of optically reading out the precise phase distribution from a single electron hologram recorded on film. There are several different methods to introduce different initial phases of a reference wave in forming electron holograms. One method is illustrated in Fig. 5.10 [5.14]. In this method, an image hologram is formed using an electron biprism [5,2]. When the illuminating electron beam is tilted as shown

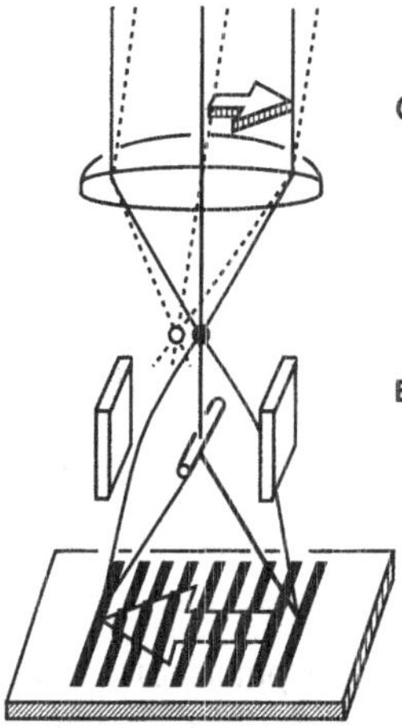

Fig. 5.10. Principle behind phase-shifting interferometry

Object

Biprism

Hologram

with the dotted lines in the figure, the interference fringes in the hologram are shifted with respect to the image. However, the image of the object does not move since the image is in focus and therefore all the electrons starting from a point in the object plane are focused onto another point in the image plane. In this way, many different holograms of the same object are formed by tilting the incident electron beam. From such holograms, the phase distribution can numerically be determined to within a precision of $\lambda/200$.

5.2.3 Three-Dimensional Image Reconstruction

When an electron beam is incident onto an object, the phase of the electron beam is modulated by electromagnetic fields, or more exactly, by electromagnetic potentials, which all will be discussed in Chap. 6. Therefore, an interference micrograph displays the phase shift of the electron beam caused by electromagnetic potentials in passing through the object. Especially for a nonmagnetic electric object which has the three-dimensional distribution $\Delta V(x,y,z)$ of electric potentials, the electron phase distribution in the micrograph indicates a two-dimensional projection. If we do not want to obtain a two-dimensional projection but the three-dimensional distribution of an electric potential, we have first to measure the projections from all directions. Then, the three-dimensional electric field can be numerically calculated from these projections, taking advantage of the **Computer Tomography** (CT) technique developed for X rays. The principles can be outlined as follows:

A parallel electron beam is incident onto an object along a certain direction and the two-dimensional phase distribution is obtained in the interference micrograph. Let us consider the xz plane in which the direction

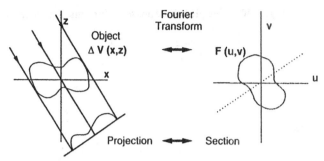

Fig. 5.11. Principle behind computer tomography

of the incident electron-beam exists. The one-dimensional phase distribution is proportional to the projection of $\Delta V(x,z)$, i.e., $\int \Delta V(x,z)\,ds$ (Fig. 5.11). There is a mathematical theorem according to which the Fourier transform of the one-dimensional distribution of the projection $\int \Delta V(x,z) \times ds$ is equal to the corresponding line section of the two-dimensional Fourier transform of the electric potential $\Delta V(x,z)$. Therefore, the Fourier transform $F(u,v)$ of the two-dimentional electric potential $\Delta V(x,z)$ can be determined by putting together all the projections measured from interference micrographs. $\Delta V(x,z)$ can be obtained by the Fourier transform of $F(u,v)$. The three-dimensional electric potentials can finally be reconstructed by piling up these planes.

While, the three-dimensional reconstruction of a magnetic field is not so easy, since magnetic fields cannot be represented by scalars like electric potentials, but by vectors. In order to consider this we need to know how an electron wave interacts with a magnetic field. Therefore, this will be discussed together with applications in Chap. 7. There, we shall introduce this technique to reconstruct both the three-dimensional shape of fine particles [5.15] and the three-dimensional distribution of the magnetic fields [5.16].

5.2.4 Real-Time Observations

Optical image reconstruction from a hologram recorded on photographic film is simple but time-consuming because of the wet processing needed to develop the film [5.8]. On-line or real-time processing techniques by means of computers and optical techniques are therefore developed. For example, electron holograms are observed via a television system attached to the electron microscope. An image can be reconstructed from the hologram by carrying out Fourier transformations twice with a computer and by numerically correcting for the spherical aberration. Both the phase and amplitude images can be determined. Images can be obtained in less than a few min-

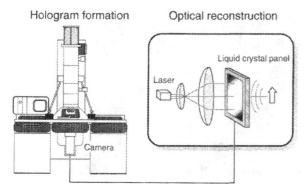

Fig. 5.12. Real-time electron holography using a liquid-crystal panel

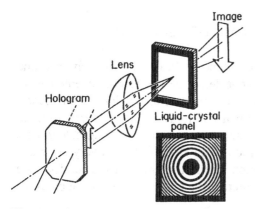

Fig. 5.13. Optical reconstruction system using a liquid-crystal panel

utes (the time depends on the performance of the computer), but it is not yet possible to achieve images in real time.

In a recently designed real-time reconstruction method [5.17], the electron hologram is detected with a TV camera and as a video signal transferred to a **liquid-crystal panel** (Fig. 5.12). Since this panel acts as a phase hologram, illuminating it with a laser beam produces an image or an interference micrograph in real time.

A liquid-crystal panel can also be employed to process a reconstructed image. For example, since the lens aberrations – including image defocusing – are equivalent to adding the corresponding phase distributions in the Fourier transform of an image, a liquid-crystal panel located at the back focal plane of an image-forming lens can be used as a phase shifter to adjust for the abberation effects (Fig. 5.13). Because the complete wave is reconstructed from the hologram, various aberration-corrected images or images under arbitrary defocusing (Fig. 5.14) can be obtained from a single hologram merely by changing the phase distribution [5.18].

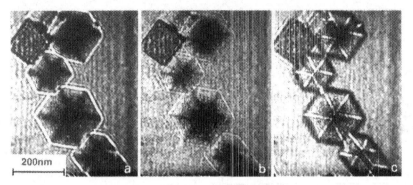

Fig. 5.14a-c. Focus series of reconstructed images of MgO particles: (**a**) $\Delta f = 10\ \mu m$, (**b**) $\Delta f = 0\ \mu m$, and (**c**) $\Delta f = 10\ \mu m$

5.2.5 Image Restoration by Aberration Compensation

The resolution of an electron microscope is limited by the spherical and chromatic aberrations of its objective lens. Dennis Gabor invented holography in order to compensate for the spherical aberration. A monochromatic electron beam has been indispensible to avoid image blurring due to the chromatic aberration, which cannot be corrected by using holography. The effect of the spherical aberration in an electron lens is very large and greatly influences the high-resolution electron micrographs. Therefore, we must always be careful when interpreting the images – even when plausible fine structure is observed.

Before going into details of the aberration correction, lens aberrations are briefly reviewed here. Through an ideal lens, all the rays starting from a point object are focused onto a single image point (Fig. 5.15a). However, no actual lens is ideal. For example, a ray starting from point O on the axis and passing through a peripheral part of a convex optical lens with spherical surfaces crosses the axis at point F' nearer to the lens than the Gaussian image point F (Fig. 5.15b). The amount of the **spherical aberration** of the lens is usually represented by the distance $\overline{FF'}$ when the object point O is infinitely far from the lens. The distance $\overline{FF'}$ is a function of α, and can be expanded in power of α. In the case of small α,

$$\overline{FF'} = C_s \alpha^2 \ . \tag{5.3}$$

The linear term has vanished from the focusing condition, and higher-order terms are neglected.

When the lens action is expressed in optical-wave terms, a divergent spherical wave from the point object O is converted into a spherical wave converging into the image point F in an ideal lens (Fig. 5.15c). That is, the

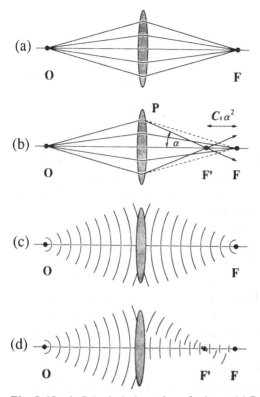

Fig. 5.15a–d. Spherical aberration of a lens: (**a**) Ray diagram with an ideal lens, (**b**) ray diagram with a lens having spherical aberration, (**c**) wave-front diagram with an ideal lens, and (**d**) wave-front diagram with a lens having spherical aberration

optical path lengths along all the rays are equal. This ensures that a perfect image can be formed through the lens with regard to its amplitude and phase.

The effect of spherical aberration can be expressed by a phase-difference between the two rays, one along the axis and the other tilted toward the axis (Fig. 5.15d). The path difference between the two rays with and without spherical aberration (solid and dotted lines in Fig. 5.15b) starting from O and arriving at the image plane is $C_s \alpha^4/4$, as determined from a simple geometrical calculation [5.19]. Therefore, the phase shift of a ray crossing the axis under the angle α relative to a ray along the axis is given by the following **wave-front aberration** $W(\alpha)$

$$W(\alpha) = \frac{2\pi}{\lambda}\frac{1}{4}C_s \alpha^4 \ . \tag{5.4}$$

Consider a concrete example for the effect of spherical aberration: the image of a crystalline object formed with a Bragg-reflected electron beam is located at a position different from that of the image formed with a transmitted electron beam under in-focus conditions (Fig.5.15b). In this way, the point resolution of an electron microscope is completely determined by the aberrations of its objective lens. This resolution limit is two orders of magnitude larger than the wavelength-determined fundamental limit. If this spherical aberration can be compensated, it would be possible to obtain a resolution of less than 100 pm.

Correction of coma aberration of an electron lens was experimentally carried out in the reconstruction stage of electron holography by *Saxon* [5.20]. Spherical aberration was compensated by *Tonomura* et al. [5.21] in a way that will be explained in detail below. Images that are optically reconstructed using electron holography are subjected to the aberrations of the objective lens in an electron microscope since the spherical aberration disturbs only the object wave and not the reference wave. This is because the reference wave passes only through the center of the objective lens.

There are several techniques for correcting a spherical aberration in the optical reconstruction stage. Examples are: (i) placing an aberration-correction phase plate at the back focal plane of an image-forming lens, as *Gabor* proposed [5.22]; (ii) using a concave optical lens to apply a negative aberration to the image; and (iii) using the aberration inherent in the imaging method of holography to correct the image.

Tonomura et al. [5.21] used method (ii), which was theoretically proposed by *Weingärtner* et al. [5.23]. In their actual experiment, however, a conjugate image, which has a negative spherical aberration, was restored onto an aberration-free image by using a convex lens instead of a concave lens, simply because the optical reconstruction system becomes simpler when a convex lens is used.

An optical system for the correction of spherical aberration is depicted in Fig.5.16. From two reconstructed images, an aperture selects only the conjugate image and an aberration-free image is obtained by using a convex lens to correct for the spherical aberration of this image. The magnitude of the spherical aberration to be corrected is given by

$$C_s' = - m^4 (\lambda/\lambda')^3 C_s \qquad (5.5)$$

where C_s is the spherical aberration constant of the objective lens in the electron microscope. The constant is defined in the following way: consider an electron beam which is parallel to the optic axis, incident on the perpendicular part of the lens, and focused into a point. However, due to spherical aberration of the lens, the distance between the lens and the focal point is given not by f but $f - C_s \alpha^2$ (f being the focal length, and α the deflection angle of an electron beam incident parallel to the optical axis at the lens). The value C_s' is the aberration constant for the optically reconstructed con-

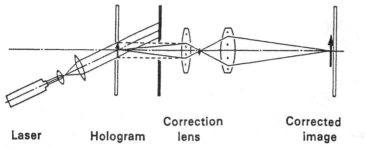

Fig. 5.16. Optical reconstruction system for correction of spherical aberration

jugate image. The value of C_s in this experiment was 1.7 mm, and 240 m that of C_s'. This C_s' value is so large that an image demagnification of about 1/6 is necessary in the correction lens. The value of C_s' can finely be adjusted by changing the demagnification rate. The experimental results will be introduced in Sect. 8.2.

The spherical aberration can also be compensated for by applying a numerical method. The principle behind this process is completely similar to the optical approach: In the first place, the Fourier transform of a hologram is calculated, which, in the optical reconstruction case, is performed at the back focal plane of a lens (being the correction lens in Fig. 5.16). Then, the effect of the spherical aberration is corrected. That is, the wavefront aberration (5.4) introduced by the electron lens is numerically subtracted from the Fourier-transform of the hologram. Finally, the aberration-free image can be obtained by calculating its Fourier transform once more.

Other methods than electron holography for correcting the lens aberrations did not reach the point where images with better resolution could actually be obtained. Recently, however, *Haider* et al. [5.24] developed an electron-optical system using two hexapoles for correcting the aberrations of the object lens in an electron microscope, and improved the image resolution from 0.24 to 0.14 nm.

5.2.6 Micro-Area Electron Diffraction

The micro-area electron diffraction technique has been developed and employed for micro-analysis purposes [5.25]. However, specimen contamination has become more and more serious as the selected areas have become smaller. It is also difficult to get diffraction patterns from an area whose shape is arbitrary. Electron diffractions from an arbitrarily small area can be obtained optically from the hologram, since all the information of an electron wave can be reconstructed from it. Neither contamination nor radiation damage is produced during examination.

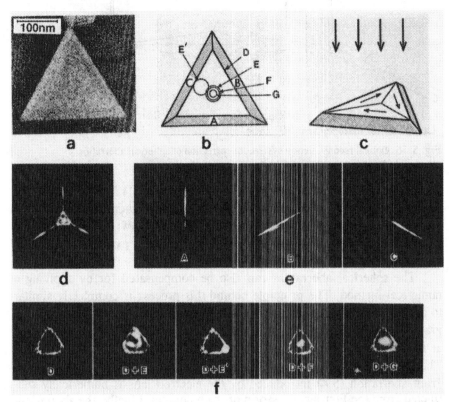

Fig. 5.17a–f. Micro-area electron diffraction patterns from various regions in a cobalt particle: (a) Reconstructed image, (b) selected areas, (c) magnetic domain structure, (d) diffraction pattern from the particle, (e) diffraction patterns from peripheral regions, and (f) diffraction patterns from central regions

An example of a triangular, fine cobalt particle is exhibited in Fig. 5.17a-f, and the following discussion refers to this figure. Low-angle electron diffraction patterns from various areas inside a fine cobalt particle are shown. No other information except the outer shape can be obtained from the electron micrograph (a). The low-angle electron diffraction (d) from the whole particle consists of three streaks and a central triangle. The three streaks could demonstrate by micro-area diffraction shown in (e) to come from the peripheral regions A, B and C in (b). The peripheral regions are wedge-shaped prisms, and they deflect the electron beam in three directions. It could also be concluded from the diffraction patterns (f) that the central triangular structure in the diffraction pattern (d) arises from the central triangular region inside the particle. Since the thickness is uniform in this region, the diffraction structure is caused by the magnetic-domain structure in the particle. The electron-diffraction patterns from small areas inside this region are presented in (f). The radii of the selected areas designated E (E'), F and G in (b) are 20, 15 and 10 nm, respectively.

Again referring to Fig. 5.17b, the diffraction pattern from area D consists of three spots and the streaks connecting them, each spot corresponds to one of the three domains shown in (c). When the selected area is smaller (like area E) the triangle is smaller, but three spots are still seen. This means that the in-plane magnetization components become smaller in the center. Diffraction patterns from the smaller central areas F and G are each composed of only one spot. If the area E is moved to one of the three domains (E'), a single spot appears at the vertex of the triangle. These investigations prove that magnetic lines of force rotate in circles inside the particle and stand up in the particle's center with a radius of 15 nm.

Ordinary electron-diffraction patterns and not low-angle diffraction patterns can be obtained from high-resolution electron holograms, as exemplified in Fig. 5.18. The diameters of the selected areas inside a fine gold particle are only 3 nm. The diffraction intensities obtained optically are equal to the electron-diffraction intensities when the holograms are recorded on film whose contrast γ is equal to -2. The diffraction patterns from two areas inside the particle differ in intensity: the intensity of the primary spot is much lower in the diffraction pattern obtained from the region B. This is because Bragg-reflected beams become more intense as the specimen thickness increases until the intensity of the transmitted beam disappears, as in region B where the half-spacing lattice image is seen. Contaminantion would become appreciable if such a small area selection were investigated by an electron probe. But we need only to make a hologram in an electron microscope, after which electron diffraction patterns can be obtained optically from desired regions without fear of contamination.

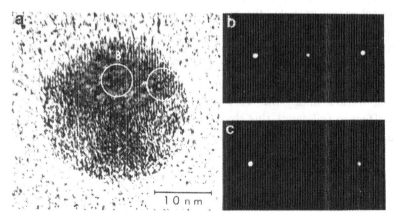

Fig. 5.18a-c. Diffraction patterns from a fine gold particle. Electron diffraction patterns are obtained from two areas 3 nm in diameter: (**a**) Reconstructed image, (**b**) diffraction pattern from region A, and (**c**) diffraction pattern from region B

6. Aharonov-Bohm Effect: The Principle Behind the Interaction of Electrons with Electromagnetic Fields

In 1959, *Aharonov* and *Bohm* presented a paper entitled "Significance of electromagnetic potentials in quantum theory" [6.1]. Its content can roughly be summarized as follows: In classical electrodynamics, potentials are merely a convenient mathematical tool for calculations concerning electromagnetic fields. The fundamental equations can always be formulated using these fields. However, in quantum mechanics, potentials cannot be eliminated from the Schrödinger equation and consequently seem to have physical significance. *Aharonov* and *Bohm* went beyond this conjecture and proposed actual electron-interference experiments. These experiments were intended to clarify how potentials affect electrons passing through field-free regions. The phenomenon these researchers described came to be called the **Aharonov-Bohm** (AB) **effect** in their honor.

The AB effect has been the subject of discussion and argument since it was first predicted theoretically [6.2]. This is not only because it contradicts common sense, but also because it is closely related to some of the most fundamental aspects of quantum mechanics, such as the physical reality of **vector potentials**. The AB effect is also important in the relation to the theory of **gauge fields** [6.3], which aims at constructing a new framework for physics. This theory extends vector potentials to non-Abelian gauge fields and proposes that the gauge fields are the most fundamental physical entity. The AB effect provides the only direct experimental evidence for this gauge principle.

The AB effect also provides the fundamental principle underlying the interaction of an electron wave with electromagnetic fields. Consequently, interference electron microscopy of magnetic samples cannot be consistently interpreted without taking the AB effect into consideration. By discussing the AB effect in full detail, this chapter offers a basis for understanding the strange behavior of the electron wave.

6.1 What is the Aharonov-Bohm Effect?

Aharonov and *Bohm* proposed two kinds of experiments: electric and magnetic. Since the electric AB effect is technically too difficult to demonstrate it experimentally, one usually means with the AB effect the magnetic type. An experimental configuration for demonstrating the magnetic AB effect is depicted in Fig.6.1. When an electric current is applied to an infinitely long solenoid, a magnetic field is produced only within the solenoid, and two electron waves starting from a single point pass around the solenoid. Even though the two waves never touch the magnetic field, a phase difference proportional to the enclosed magnetic flux is produced between them. Classical electrodynamics does not consider that electrons outside the solenoid would "know" about the magnetic flux inside.

This conclusion can be obtained in a straightforward way by solving the Schrödinger equation

$$\left[\frac{1}{2m}(-i\hbar\nabla + e\mathbf{A})^2 - eV\right]\psi = E\psi, \tag{6.1}$$

where m, $-e$, $\hbar$, ψ, $\mathbf{A}$, V are the electron mass, the electron charge, Planck's constant divided by 2π, the wave function, the vector potential, and the scalar potential, respectively. If we assume that electromagnetic fields are weak enough to influence an incident electron wave only slightly, we can solve this equation using the WKB approximation: When the wave function ψ is expressed by both the amplitude R and the phase $S/\hbar$ as follows:

$$\psi = Re^{iS/\hbar},$$

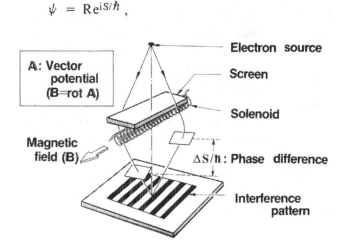

Fig. 6.1. The Aharonov-Bohm effect

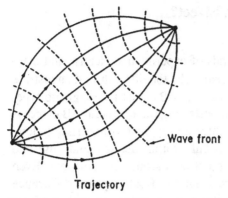

Fig. 6.2. Relation between electron trajectories and wave fronts

then (6.1) is replaced by the following two equations:

$$(\text{grad} S + e\mathbf{A})^2 - 2me V = (\hbar^2 \Delta R/R) \tag{6.3}$$

and, using $\text{div}\mathbf{A} = 0$,

$$\text{div}[R^2(\text{grad} S + e\mathbf{A})] = 0 . \tag{6.4}$$

If the fractional change of the potential V, i.e., $\lambda |\text{grad} V|/|V|$ at the distance λ, λ being the electron wavelength, is much smaller than 1 – and similarly for the vector potential $\mathbf{A}$ – then (6.3) becomes

$$m\mathbf{v} = \text{grad} S + e\mathbf{A} , \tag{6.5}$$

where $\mathbf{v}$ is the electron velocity and $|m\mathbf{v}| = (2meV)^{1/2}$.

This equation relates the phase of the electron wave function to the electron velocity. For pure electrostatic fields ($\mathbf{A} = 0$), the situation is very simple. As shown in Fig. 6.2, the electron wave front is determined by the surface perpendicular to the electron trajectory. However, for magnetic fields the situation becomes more complicated. The wave front is given by the surface perpendicular not to the electron trajectory but to the $(m\mathbf{v}-e\mathbf{A})$ line. Since the gauge freedom prevents the vector potential $\mathbf{A}$ from being defined uniquely, we are led to the strange conclusion that the wave front is also not determined uniquely. However, this does not lead to inconsistencies since one can never observe the wave front itself. But what one can observe is a phase difference (i.e., a difference in the number of wave fronts) between two electron trajectories starting from one point, tracing different paths, and ending at another point.

When a classical electron trajectory is known, the phase of the electron wave function can be derived in the non-relativistic approximation:

$$\frac{S}{\hbar} = \frac{1}{\hbar}\int (m\mathbf{v} - e\mathbf{A})\cdot d\mathbf{s} = \frac{1}{\hbar}\int \left[\sqrt{2meV} - e\hat{\mathbf{t}}\cdot\mathbf{A}\right]ds, \quad (6.6)$$

where the integral is carried out along an electron trajectory, and $\hat{\mathbf{t}}$ is the unit tangent vector of the electron trajectory. The phase difference between the two electron beams shown in Fig. 6.1 can now be obtained using (6.6):

$$\Delta S/\hbar = -(e/\hbar)\oint \mathbf{A}\cdot d\mathbf{s} = -(e/\hbar)\int \mathbf{B}\cdot d\mathbf{S}. \quad (6.7)$$

The first integral is carried out along a closed path along two electron trajectories, and the second integral is performed over the surface enclosed by the two paths. Even when there are no electromagnetic fields and the two electron beams travel along paths that are of an equal length, an observable effect is produced. The following section will use the double-slit experiment to demonstrate how inconsistent the AB effect is from a classical viewpoint.

6.2 Unusual Features of the Aharonov-Bohm Effect: Modified Double-Slit Experiments

A standard double-slit experiment is illustrated in Fig. 6.3a. The interference pattern shows its peak intensity at the center of the screen because the two beams are in phase at that location. This experiment is essentially the same as Young's experiment with light. What happens when we place a thin ferromagnetic film behind the slits? The film is assumed to be magnetized and to have negligible leakage flux. First, let us consider two sheets of ferromagnetic film placed on a plane located behind the slits, as depicted in Fig. 6.3b. There is a gap between the sheets corresponding to the space between the slits, but the film covers the portions of the plane through which the beams pass. Since the magnetic field deflects both beams, we might expect that the interference pattern is displaced as a whole. This would mean that the pattern would not necessarily have a peak intensity at the center. The fact is that the peak intensity does remain at the center because the relative phase shift between the two beams is determined by the enclosed magnetic flux, irrespective of the electron-beam deflection.

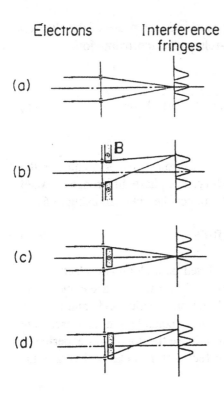

Fig. 6.3a-d. Double-slit experiments with electrons. Interference patterns are determined by magnetic flux enclosed by two-electron beams passing through slits irrespective of whether electrons are deflected by a magnetic field. Cases (**a**) and (**b**) are therefore equivalent, as are (**c**) and (**d**). According to classical electrodynamics, cases (**a**) and (**c**) are equivalent, as are (**b**) and (**d**)

The double-slit experiment sketched in Fig. 6.3c illustrates the AB effect. Although the electron beams do not pass through the magnetic flux, a phase difference of $e\Phi/\hbar$ is produced between them, and the interference fringes are displaced. Furthermore, even when the electron beams actually cross the magnetic film at its edges and are deflected (Fig. 6.3d), the interference pattern is not changed.

In classical mechanics, only the forces acting on the electrons affect their behavior. Consequently, cases (a) and (c) are equivalent. Similarly, cases (b) and (d) are also equivalent. However, in quantum mechanics the concept of forces gives way to the concept of electromagnetic potentials, and the potentials shift the electron phase. According to quantum mechanics, cases (a) and (b) are equivalent, as are (c) and (d).

We now arrive at an important question. Why does an electron notice the existence of the hidden magnetic field even though it never crosses the magnetic field? *Aharonov* and *Bohm* argued this problem as follows: In classical mechanics, electromagnetic potentials were only a mathematical aid. In quantum mechanics, however, this is not the case. *Aharonov* and *Bohm* thought that this effect could not be interpreted merely as a result of the local interaction of an electron with the electromagnetic fields. If the local interaction principle required by relativity is adhered to, there is no choice but to consider the potentials to be physical entities that are more fundamental than the fields.

6.3 The History of Vector Potentials

The physical meaning of vector potentials is not easy to understand. For this reason or others, vector potentials have been an object of discussion for the past century in the main stream of physics. The history of vector potentials is briefly outlined here on the basis of an article by *Yang* [6.4].

In 1831 M. Faraday discovered **electromagnetic induction**: When a magnet approaches a metal ring, such an electric current is induced that the magnetic flux passing through the ring may remain the same. Faraday thought that the ring under the application of a magnetic field must be in some special state called **electrotonic state** and that a current was induced only when the state changed, although he did not know what the electrotonic state was. Faraday afterwards found that he could dispense with this concept by means of cutting of magnetic lines of force across the ring. However, thereafter the idea of the electrotonic state had come up in mind again and again [6.5].

In 1856, J.C. Maxwell considered that the electrotonic state could be described by the vector potential **A**. The vector potential had first been introduced by W. Thomson who penetrated the vortical nature of magnetic lines of force (**B** = curl**A**) from the Faraday rotation. Maxwell succeeded in formulating the theory of an electrotonic state using **A**: **B** = curl**A** and **E** = $-\partial \mathbf{A}/\partial t$. These equations indicates that electric and magnetic fields are interrelated through **A**. That is, **B** is produced when a vortex exists in the flow of **A**, and **E** is produced when **A** changes with time. Maxwell conceived of **A** as a physical quantity, the "**electro-kinetic momentum**" [6.6], since the time derivative of **A** is a force exerted on a unit electric charge just as the time derivative of a momentum is a force.

Thirty years after Maxwell, O. Heaviside and H. Hertz discarded **A** in **Maxwell's equations** asserting that **A** had no physical meaning. The simplified Maxwell equations without **A** have since been widely accepted.

In the 20th century, the two big developments in physical theories were borne out of principles behind Maxwell's equations: A. Einstein constructed the concept of relativity from the constancy of the light velocity in any coordinate. The fundamental idea towards the theory of gauge fields was advanced by H. Weyl. He tried to realize Einstein's dream of unifying gravitational and electromagnetic forces by introducing the vector potential **A** to Einstein's theory of gravity, the general relativity. According to his theory, the gauge of length becomes different at each point in space-time due to the existence of **A**. Thus, he intended to unify electromagnetism and gravity as structures of space-time.

A. Einstein, though impressed with Weyl's concept, objected to it saying "it cannot be physics since a 1-m long stick can become 2-m long after making a round trip" (for a more detailed discussion see [6.7]).

Altough Weyl's attempt was unsuccessful, his way of thinking was succeeded by theories of gauge fields: In 1954, *Yang* and *Mills* made the first step toward unifying electromagnetic and nuclear forces [6.8]. The gauges of length in Weyl's theory were replaced by the references of the phase of the electron wave function. Until that time, new nuclear forces had been discovered and quantum mechanics had also been established. Vector potentials are called **gauge fields** and regarded as the most fundamental physical quantity in these theories.

Einstein's objection on these new theories corresponds to the AB effect: The phase of an electron can change after making a round. The AB effect thus became important in the theory of gauge fields.

6.4 Fiber-Bundle Description of the Aharonov-Bohm Effect

The physical significance of the AB effect was more fully appreciated in the late 1970s. The theory of gauge fields, proposed in 1954 by *Yang* and *Mills* [6.8] and in 1956 by *Utiyama* [6.9], was taking on a greater significance. After the successful unification of electromagnetism and the weak force by *Weinberg* [6.10] and *Salam* [6.11], the theory of gauge fields came to be regarded as the most probable candidate for a unified theory of all interactions in nature. In 1975, *Wu* and *Yang* [6.3] formulated a complete description of electromagnetism that focused on a new physical variable called a **non-integrable** (path-dependent) **phase factor**. By using the mathematical theory of fiber bundles, they generalized this formulation to a non-Abelian gauge field. In their formulation, the non-integrable phase factor corresponds to the parallel transport on a curved surface. In this context, the AB effect relates to a self-evident geometrical theorem, and it is considered to be an experimental manifestation of the non-integrable phase factor.

Concerning the most fundamental physical quantity in electromagnetism, *Wu* and *Yang* [6.3] pointed to a physical quantity that has a one-to-one correspondence with what we can experimentally observe in electromagnetism, but that is neither the field strength $F_{\mu\nu}$ nor the vector potential A_μ. Instead they considered

$$\exp\left(\frac{-ie}{\hbar}\oint A_\mu\,dx_\mu\right) = \exp\left[\frac{-ie}{\hbar}(\oint \mathbf{A}\cdot d\mathbf{s} - \oint V\,dt)\right]. \qquad (6.8)$$

By comparing this equation with (6.7) we may conclude that the phase factor is exactly what we can observe in electron-interference experiments. The integral $\oint A_\mu\,dx_\mu$ is carried out along a closed loop determined by two paths. However, when the integral must be performed not along a loop but along a path connecting the two points P and Q, then the quantity is not only a function of the points P and Q but it is path dependent. It is described by the following non-integrable phase factor

$$\exp\left(\frac{-ie}{\hbar}\int_P^Q A_\mu\,dx_\mu\right), \qquad (6.9)$$

being pictured in a practical AB-effect arrangement (Fig. 6.4), is a complex number. Its value depends on the two points P and Q, and it is different for each topologically different path. Paths *1* and *2* pass on the left and right sides of the solenoid, and path *3* goes around the solenoid once in a clockwise direction: the phase factors for these three paths assume different values.

The preceeding discussion clearly reveals that the phase factor reflects the topological nature of the path it belongs to. However, this non-integr-

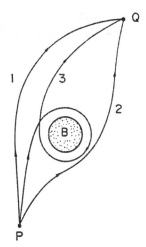

Topological Paths around a Solenoid

Fig. 6.4. Phase factors having different values along different paths

able phase factor is not observable because it is impossible to measure the phase shift along a path from P to Q. The only way to observe the phase difference is to look at the interference pattern created by the phase difference relative to another path from P to Q. What can actually be observed is then described by the phase factor whereby the integral is performed along the closed loop.

This situation were fully expressed by *Wu* and *Yang* [6.3] as follows:

- The local field strength $F_{\mu\nu}$ *under-describes* electromagnetism. This is proven by the AB effect.
- The phase $(ie/\hbar) \oint A_\mu \, dx_\mu$ *over-describes* electromagnetism, since no experiment can distinguish between cases where the values differ only by multiples of 2π.
- Electromagnetism is *completely described* by the phase factor

    $\exp[-(ie/\hbar) \oint A_\mu \, dx_\mu]$.

This phase factor contains the necessary and sufficient observable information about electromagnetism.

Amazingly, all electromagnetic phenomena, with all their various aspects, can then be expressed by the phrase *electromagnetism is the gauge-invariant manifestation of a non-integrable phase factor*. Furthermore, the formulation of electromagnetism using a non-integrable phase factor was generalized to **non-Abelian gauge fields**. The significance of the phase factor increases there, because the field strength underdescribes the gauge field even in a simply-connected region. This formulation was shown to be expressed in purely geometrical terms by the **fiber-bundle theory** [6.3]. This theory describes the geometry of manifolds. A simple example providing a glimpse of this theory is depicted in Fig.6.5a. If a two-dimensional human being that lives on the earth wants to know the earth's curvature, he can do so by the parallel transport of an arrow. He moves from the

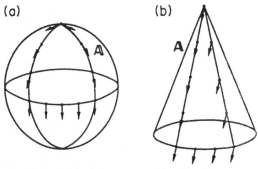

Fig. 6.5a,b. Parallel transport of a direction vector: (**a**) Spherical surface, and (**b**) conical surface

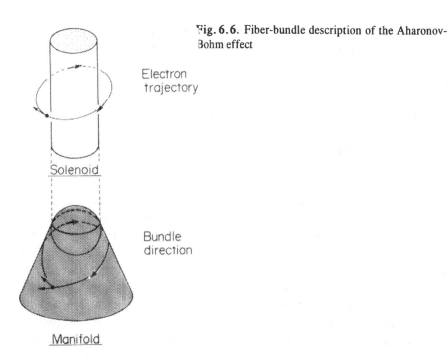

Fig. 6.6. Fiber-bundle description of the Aharonov-Bohm effect

north pole to the equator, then moves 90° along the equator, and finally returns to the north pole. What happens to his arrow? Although the arrow comes back to the orginal position after the parallel transport, its direction does not return to the original direction, but changes by 90°. This change in direction indicates that the earth's surface is not flat. The direction of the arrow does not change after parallel transport along the closed path on a conical surface (Fig.6.5b) since the conical surface is essentially flat. This can easily be seen by cutting and unrolling the surface.

Thus, we can find the curvature of the surface by the operation *parallel transport*. In the context of the theory of gauge fields, a non-integrable phase factor corresponds to parallel transport in a fiber bundle. The electromagnetic field and vector potential, respectively, relate to the curvature of the bundle and a principal connection. The electron phase shift corresponds to the change in the arrow direction. The AB effect can be described by using parallel transport on a curved surface [6.12], which consists of a conical surface capped with a sphere (Fig.6.6). The curvature is nonzero only on the spherical surface, which corresponds to the interior of the solenoid where a nonzero magnetic field exists. Although the arrow direction seems to return to the original direction by parallel transport along any closed path on the conical surface where the curvature vanishes ($\mathbf{B} = 0$, Fig. 6.5b), a difference in arrow direction is produced when the closed path encloses the cone vertex (Fig.6.6). This can easily be understood by parallel transport of the arrow on the plane onto which the conical surface is slit longitudinally and unrolled; it precisely simulates the AB effect.

6.5 Early Experiments and Controversy

6.5.1 Early Experiments

The AB effect was experimentally tested just after it was predicted, and the first result was reported by *Chambers* [6.13]. He used a thin iron whisker in the shadow of the electron biprism filament, and detected a phase shift proportional to the enclosed magnetic flux as a biprism interference pattern. Soon after *Chambers* several other experiments were reported [6.14-16], all detecting the existence of the AB phase shift. Among the experimental results, those of *Möllenstedt* and *Bayh* [6.16] were both elegant and physically significant. A conceptual diagram of their experiment is depicted in Fig. 6.7a. It shows a collimated electron beam split by a biprism into two coherent beams. Thereafter the beams are overlapped to form an interference pattern. An extremely thin solenoid is located between the two beams. In the actual experiment, three electron biprism stages were provided so that the two beams might be kept far enough apart ($\approx 120\,\mu$m) that they would not illuminate the solenoid. In addition, a ferromagnetic yoke was provided

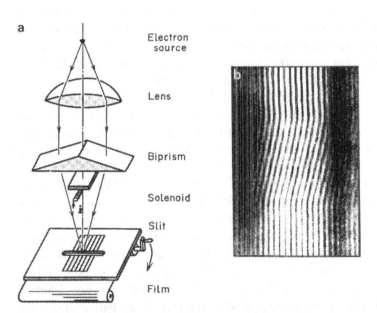

Fig. 6.7a,b. The Aharonov-Bohm effect experiment by *Möllenstedt* and *Ba h* [6.16]: (**a**) Electron-optical diagram, and (**b**) interference pattern

to remove the effect of fringing fields. When the electric current applied to the solenoid was increased, the interference fringes moved laterally while the region of the whole interference pattern remained unchanged. To record this behavior, only a part of the interference pattern was recorded on film through a slit perpendicular to the fringe direction. When the film was made to move together with the increase in magnetic flux, dynamic behavior could be caught in the photograph shown in Fig.6.7b. The slit is in the horizontal direction and the change of the interference pattern with time can be seen along the vertical axis. The fringes remain unchanged until a current is applied. When the current increases, the fringes are shifted. Even when the current stops increasing, the fringe shift persists, thus demonstrating the existence of the AB effect.

6.5.2 Nonexistence of the Aharonov-Bohm Effect

The AB effect has been controversial ever since it was predicted. However, until the later 1970s, discussions were focused on theoretical interpretations of the AB effect. Experiments were generally thought to demonstrate that the AB effect does exist. The significance of the AB effect increased greatly when it came to be regarded as experimental evidence for the theory of gauge fields. As time progressed, even the existence of the phenomenon has been questioned, and controversy flared anew.

In 1978, *Bocchieri* and *Loinger* [6.17] claimed that the AB effect did not exist. They asserted that the AB effect is actually gauge-dependent and that it is a purely mathematical concoction. They claimed that all consequences of quantum mechanics depend on field strengths and not on potentials. Although they put forward many different reasons for this assertion, only the most salient are summarized here.

a) Non-Stokesian Vector Potential

According to the analysis of *Bocchieri* and *Loinger*, a gauge function can be chosen so that the vector potential **A** completely vanishes outside an infinite solenoid: consequently there is no AB effect [6.17, 18]. The vector potential **A** in the Coulomb gauge around a solenoid with radius a that contains the magnetic flux Φ can be given by

$$A_r = A_z = 0, \quad A_\theta = \frac{\Phi}{2\pi r} \quad (r \geq a),$$

$$A_r = A_z = 0, \quad A_\theta = \frac{\Phi r}{2\pi a^2} \quad (r < a).$$

(6.10)

Here the vector potential vanishes outside the solenoid when the gauge transformation

$$\Lambda = -\frac{\Phi\theta}{2\pi} \tag{6.11}$$

is introduced. The new vector potential $\mathbf{A}' = \mathbf{A} + \text{grad}\Lambda$ then becomes

$$\mathbf{A}' = 0 \quad (r \geq a),$$
$$A'_r = A'_z = 0, \quad A'_\theta = \frac{\Phi r}{2\pi a^2} - \frac{\Phi}{2\pi r} \quad (r < a). \tag{6.12}$$

This vector potential does not satisfy **Stokes'** theorem and is called **non-Stokesian**. An electron passing outside of the solenoid is therefore unaffected by the field in the solenoid.

Home and *Sengupta* [6.19] asserted that the existence of the AB effect depends on its time-dependent aspect when an electric field circulating around the solenoid has been produced during the turning-on process of the magnetic field, and that non-Stokesian vector potentials were not excluded by presenting the different example:

$$A''_r = A''_z = 0, \quad A''_\theta = \frac{\Phi r}{2\pi a^2}. \tag{6.13}$$

b) Hydrodynamical Formulation

According to *Bocchieri* and *Loinger* [6.17], the Schrödinger equation can be replaced by a set of nonlinear differential equations called **hydrodynamical equations**, which contain only the field strengths $\mathbf{E}$ and $\mathbf{B}$. There is therefore no room for the AB effect. The hydrodynamical equations are expressed as

$$\text{div}(\rho\mathbf{v}) + \frac{\partial\rho}{\partial t} = 0, \tag{6.14}$$

where

$$\rho = \Psi\Psi^* = |R|^2, \quad \text{and} \quad m\frac{d\mathbf{v}}{dt} = -e(\mathbf{E} + \mathbf{v}\times\mathbf{B}) + \frac{\hbar^2}{2m}\text{grad}\left[\frac{\nabla^2 R}{R}\right],$$

$$\tag{6.15}$$

where $m\mathbf{v} = \text{grad} S + e\mathbf{A}$.

c) Doubts About the Validity of Early Experiments

Bocchieri et al. [6.18] also expressed doubts from experimental viewpoints about the existence of the AB effect: they claimed that the interference experiments must have been affected by leakage fields from solenoids or whiskers. They asserted that *Chamber*'s results could be fully explained by a magnetic-field leakage from the whisker.

With regard to the rather comprehensive experiment by *Möllenstedt* and *Bayh* [6.16], *Bocchieri* et al. [6.18] argued that there must be a significant magnetic field in the space between two consecutive turns of the solenoid. This field, which has a non-oscillating component as well as one that is spatially oscillating, is approximately parallel to the solenoid axis. An electron entering this field would feel a Lorentz force. The fringes would not be shifted if the electron wave did not penetrate into the solenoid, but even a small penetration would be enough to generate an appreciable fringe shift.

In 1980, *Roy* [6.20] proposed that physical effects depend only on accessible fields, and that no effect due to inaccessible fields can exist. The magnetic fields leaking from both ends of a finite solenoid in the previously reported experiments could not be interpreted as being caused by inaccessible magnetic fields. He stressed that a solenoid of finite length, even if surrounded by an impenetrable cylinder, yields a potential with asymptotic behavior that excluded effects of the inaccessible fields.

6.5.3 Dispute About the Nonexistence of the Aharonov-Bohm Effect

Several scientists argued against the claim that the AB effect would not exist. Since it is difficult to review every detail, these arguments are summarized below:

a) Non-Stokesian Vector Potentials

Ehrenberg and *Siday* [6.21] predicted the magnetic AB effect in 1949 when they formulated electron optics in terms of the **refractive index** μ whose definition was based on **Fermat's principle**:

$$\mu = 1 + \frac{\hat{t} \cdot A}{Br}, \tag{6.16}$$

where B is a uniform magnetic field, r is the radius of a circular electron trajectory in that field, and $\hat{t}$ is the unit vector along the electron path.

Ehrenberg and *Siday* stated that in order to avoid violating the validity conditions for Fermat's principle, the refractive index μ should: (i) be fixed everywhere in space once it is fixed in the neighborhood of one point, (ii) have no singularities that make the integral in Fermat's principle convergent, and (iii) have only such discontinuities as those which appear as limiting cases of μ. The vector potential must therefore satisfy Stokes' theorem. Under these restrictions, they said, the vector potential outside an infinitely long solenoid cannot, in general, vanish with the magnetic field.

After non-Stokesian vector potentials were claimed to negate the existence of the AB effect, their inadmissibility was asserted by several researchers. For example, *Bohm* and *Hiley* [6.22] made the following assertions. The two kinds of vector potentials – in the Coulomb gauge, **A** given by (6.10); in the Bocchieri and Loinger gauge, **A**' given by (6.12) – seem at first sight equivalent to each other since they are related by the gauge transformation (6.11). Here, the non-Stokesian vector potential **A**' does not fully describe the physical situation of the infinite solenoid. Rather, it describes a different situation, one in which, in addition to the original solenoid with the magnetic flux Φ, there is an infinitely thin solenoid with the magnetic flux $-\Phi$ along the central axis. This can easily be confirmed by calculating the magnetic field **B**' from **A**' in (6.12). The resultant magnetic field is

$$\mathbf{B}' = [B - \Phi \delta^{(2)}(\mathbf{r})]\hat{\mathbf{z}} \quad (r \le a), \tag{6.17}$$

where $\hat{\mathbf{z}}$ is a unit vector in the solenoid direction, and $\delta^{(2)}(\mathbf{r})$ the delta function. Thus, Stokes' theorem holds for the vector potential **A**', too. In this way, the total magnetic flux inside the solenoid vanishes for this magnetic-field distribution. This is why the AB effect cannot be derived from the gauge of *Bocchieri* and *Loinger*.

b) Hydrodynamical Formulation

Bocchieri and *Loinger* claimed that there could be no AB effect because the Schrödinger equation can be reformulated in the form of a set of hydrodynamical equations using only field strengths. *Bohm* and *Hiley* [6.22], and *Takabayasi* [6.23] asserted that in this formulation the AB effect can be produced in the form not of local interaction but of non-local interaction. To make the hydrodynamical equations (6.14, 15) equivalent to the Schrödinger equation, the following non-local equation should be added to them

$$m \oint \mathbf{v} \cdot d\mathbf{s} = nh + e \oint \mathbf{A} \cdot d\mathbf{s}. \tag{6.18}$$

The line integral is performed outside the solenoid. This equation comes from the single-value condition of the wave function, and the AB effect is produced by the second term in this equation. *Bocchieri* et al. [6.24] again stressed that the second term vanishes for the non-Stokesian vector potential given by (6.12).

c) Discussions on the Validity of Experiments

Discussions on the validity of early experiments took the form of repeated criticisms and responses, making it complicated to review each detail. These discussions appeared as if they might continue endlessly. It is therefore not surprising that after 1980 new kinds of experiments were proposed to serve as crucial tests for the existence of the AB effect.

In 1980 *Kuper* [6.25] proposed an electron-diffraction experiment using a hollow superconducting torus (Fig.6.8). A toroidal magnetic field would be trapped inside the torus of the superconductor, and the total flux would consequently be quantized as $nh/(2e)$ (n being the number of flux quanta). In this experiment an electron wave is incident on the torus, and the two partial waves, having passed inside and outside the torus, are superimposed on a screen to form a diffraction pattern. The geometry here is similar to that for a zone plate, and whether the central intensity of the diffraction pattern is bright or dark depends on whether an even or odd number of magnetic flux quanta are trapped. This results from the AB effect giving the two waves a relative phase shift of either 0 or π modulo 2π causing them to interfere either constructively or destructively at the center of the screen.

Roy [6.20] had claimed that all previous experiments could be explained as resulting from electromagnetic fields outside the solenoid. He felt that this explanation would remain valid because the ideal geometry – an in-

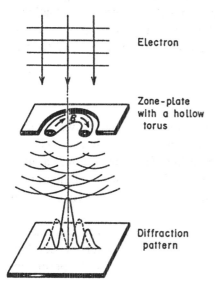

Fig. 6.8. Proposal for an experiment to decisively confirm the *Aharonov Bohm* effect by using a superconducting hollow torus [6.25]

finitely long solenoid – is unattainable. However, one approach he did not deal with was utilizing a toroidal magnetic field. In 1978 *Lyuboshitz* and *Smorodinskii* [6.26] theoretically investigated the AB effect for a toroidal solenoid and reported that the total cross section of scattering, which diverges for an infinite solenoid, becomes finite for a toroidal solenoid.

6.6 Experiments Confirming the Aharonov-Bohm Effect

Theoretical discussions may be heuristically useful, but it is the experimental evidence alone that can determine whether or not the AB effect exists. Since the electron wavelength is so short, experimental setups are required to be extremely small. The sample size, for example, must be less than 10 μm so that the whole sample can be fully contained in the spatially coherent region of the illuminating electron wave. In addition, a sample has to be specially designed so that a perfect experiment will avoid problems such as the leakage flux. By the 1980s, technological progress, particularly in areas such as **microlithography** techniques developed for manufacturing semiconductor devices, opened up a window of opportunity for fabricating the required tiny and complicated samples.

6.6.1 An Experiment Using Transparent Toroidal Magnets

As *Kuper* [6.25] stressed, the ideal geometry of an infinitely long solenoid is experimentally unattainable. However, an ideal geometry can actually be achieved by using a toroidal solenoid. An experiment was carried out for the first time in 1982 using transparent toroidal ferromagnets [6.27], it is described below:

a) Sample Preparation

Tiny ferromagnetic samples with a square toroidal geometry were fabricated photolithographically. Films of 40 nm thick permalloy (80% Ni and 20% Fe) were prepared by vacuum evaporation. The substrate was a glass plate covered by evaporation with a thin film of NaCl. A photolithographic process was then employed to cut toroidal samples from the permalloy film (Fig. 6.9).

A photoresist was first coated over the permalloy surface, and the toroidal photomask pattern was transferred to the resist. After the development process, only the resist in a toroidal shape remained on the permalloy

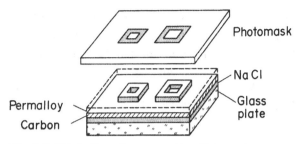

Fig. 6.9. Fabrication of transparent toroidal samples

film. The film, except for the area covered with the resist, was then ion-milled. The photoresist left on the permalloy toroids was removed by fixing. The toroids were floated off on a water surface and were applied to a carbon film about 10 nm thickness. The magnetic flux flowing inside such toroids could be estimated from the width of the toroids to range from 3(h/e) to 10(h/e). An electron holographic interference microscope was used to measure leakage fields from the toroidal samples and the only samples selected were those whose leakage flux was too small to influence the conclusions drawn from the experiment.

Two typical interference micrographs of the toroidal samples with and without leakage flux are depicted in Fig. 6.10. As will be explained in Chap. 7, contour fringes in an electron interference micrograph follow in-plane magnetic lines of force in flux units of h/e [6.28, 29]. Since these micrographs are phase-amplified twice, a constant magnetic flux of h/(2e) flows between two adjacent contour fringes. In micrograph (a), almost all of the

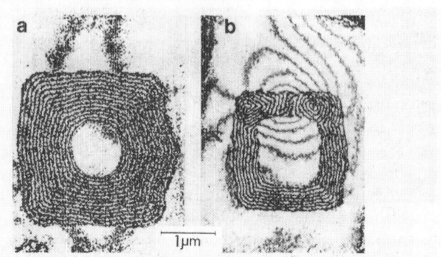

Fig. 6.10a,b. Interference micrographs of toroidal samples (phase amplification: ×2): (**a**) Closed magnetic circuit, and (**b**) open magnetic circuit

flux is confined within the magnet: it can be seen that only 1/15 of the total flux leaks outside. This means the leakage can effect the phase of an incident electron beam by as little as π. However, in micrograph (b), all of the flux leaks outside the toroid, perhaps because of imperfections in the toroidal shape. Only samples like that shown in Fig. 6.10a, which were measured to be free from leakage flux, were selected for the experiment.

ι) **Experimental Results**

The phase distribution of an electron beam transmitted through such a toroidal sample was observed as an interferogram [6.27]. When there is no relative phase shift between an electron beam passing outside the toroid and one passing through the hole in the toroid, interference fringes inside and outside the toroid must lie along the same straight lines. The interferogram shown in Fig. 6.11a clearly indicates the existence of a relative phase shift of about 12π. Such a phase shift might be attributed to a change in specimen thickness. This possibility can be excluded when the interferogram of another sample (Fig. 6.11b) reveals that the wave-front displacement is reversed. This cannot be explained by a thickness change and must be due to the AB effect when the magnetization direction in the toroid is reversed.

The AB phase shift was also confirmed quantitatively. Since the thickness of the toroid is 40 nm and the magnetic induction is equal to 0.95 Wb/m^2, the magnetic flux flowing through the toroid whose interferogram is shown in Fig. 6.11a (and whose average width is 640 nm) is $2.4 \cdot 10^{-14}$ Wb or 6.0 (h/e). This flux is consistent with the observed 12π phase shift. The

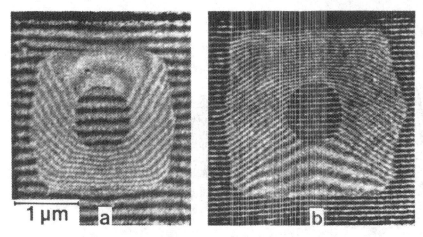

Fig. 6.11a,b. Experimental results showing a relative phase shift between beams passing through the hole in a toroid and outside the toroid: (**a**) Clockwise rotation of magnetization, and (**b**) counter-clockwise rotation of magnetization

toroid whose interferogram is depicted in Fig.6.11b (toroid width: 860 nm) has a flux of 8.0 (h/e), which is also consistent with the fringe shift.

This quantitative agreement is characteristic of the AB effect. That is, the amount of the phase-shift value is completely determined by the enclosed magnetic flux and does not depend on any other experimental conditions, such as the electron energy. It should be noted in passing that the amount of a phase shift due to a thickness change depends on the electron energy. These results support the existence of the AB effect even under conditions where there is no leakage flux.

c) Discussions of the Validity of the Experiment

The validity of this experiment as a test [6.27] for the AB effect was questioned by *Bocchieri* et al. [6.30] as well as by *Miyazawa* [6.31]. *Bocchieri* et al. [6.30] asserted that the observed effect could be explained by taking into account the action of the Lorentz force on the portion of the electron wave that penetrated into the magnet. They pointed out that the penetration generally induces a sudden adjustment of the whole electron wave. They claimed that this adjustment precisely simulated the AB effect and that this experiment therefore provided no proof for the AB effect. They remarked that when designing an experiment for testing the AB effect, a key feature includes all the electron waves that traverse regions free of the magnetic field.

Miyazawa [6.31] also argued that the experiment reported by *Tonomura* et al. [6.27] did not effectively test for the existence of the AB effect. He claimed that in an experimental test for the AB effect, the wave function of an incident electron beam must be zero at all the boundaries of the magnet. *Miyazawa* was confident that if such boundary conditions were to be satisfied in a new experiment, the so-called **AB effect** would vanish. His reasoning was as follows: He first discussed whether or not the wave function is single-valued. He said that multi-valued wave functions cannot be excluded *a priori*, as several scientists such as, for example, *Schrödinger* [6.32] and *Pauli* [6.33] considered such possibility seriously. After giving up the postulate of single-valuedness regarding the AB effect for bound-state electrons orbiting a magnetic string, *Miyazawa* adopted the minimum-energy principle. He concluded that when the electron wave function and the magnetic field overlap, the wave function is single-valued, producing the so-called AB effect. However, when the overlap decreases, his interpretation was that the wave function becomes multi-valued, causing the disappearance of any AB effect. That is, the electron makes a transition from one state to the other when this overlap changes.

6.6.2 An Experiment Using Toroidal Magnets Covered with a Superconducting Film

The experiments using transparent toroidal ferromagnets did not convince those who doubted the existence of the AB effect. On the one hand, they asserted that the detected phase shift could be interpreted as a result of the Lorentz force on the electrons which partially penetrated the magnet, and on the other hand, they said the leakage flux must be reduced to zero for a true test. An experiment that satisfied the conditions for a decisive experimental test of the AB effect was carried out in 1986 [6.34-36] and is described below.

a) Sample Preparation

To preclude the electron penetration effect, an additional metal layer had to completely shield the toroidal magnet from an incident electron beam (Fig.6.12). In this way, the sample becomes free from electron penetration. However, the consequence is that the absolute phase difference cannot be measured unless it is modulo 2π. Another problem is that although an electron beam cannot penetrate the magnet, a small amount of leakage flux from the magnet might influence the electron beam. To prevent this possibility, the shielding metal layer was made of superconducting material. The Meissner effect prevents magnetic fields from passing through a superconducting layer. Consequently, electron holography showed that no magnetic fields leaked outside the fabricated toroidal sample.

It thus became possible to look for the AB effect by using this sample under conditions where there was no overlap between the magnetic field and an electron beam. A tiny toroidal magnet less than 10 μm in diameter had to be completely covered by a superconductor, without even the slightest gap. The thickness of the superconducting layer had to be larger than

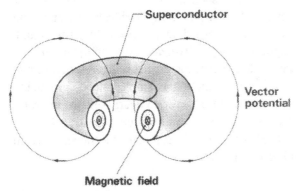

Fig. 6.12. A toroidal sample with no overlap between a magnetic field and an electron beam

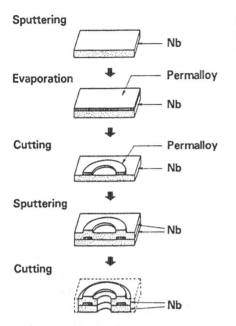

Fig. 6.13. Photolithographic process for fabricating a toroidal ferromagnet covered with a superconductor

the penetration depth. Such samples could actually be fabricated by using advanced photolithographic techniques. The fabrication process is complicated, so only its principle aspects are represented in Fig. 6.13.

A 20-nm thick permalloy film was prepared by vacuum evaporation on a silicon wafer covered with a 250-nm thick niobium film. After the toroidal shape was cut out of the permalloy film, a 300 nm niobium film was

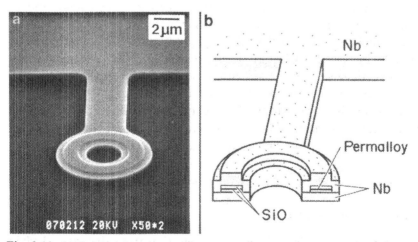

Fig. 6.14a,b. Toroidal ferromagnet covered with a superconductor: (a) Scanning electron micrograph, and (b) cross-sectional diagram. The bridge between the toroidal sample and the Nb plate is for cooling the sample

71

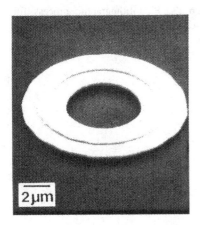

Fig. 6.15. Scanning electron micrograph of a toroidal ferromagnet without a bridge

sputtered on its surface. Then, the toroidal sample was cut so that the permalloy toroid was completely covered with the niobium layer. Finally a copper film 50 to 200 nm thick was evaporated on all of its surfaces; this film prevented electron-beam penetration into the magnet and kept the sample from experiencing charge-up and contact-potential effects. In a scanning electron micrograph of the sample (Fig.6.14), one can see a bridge between the toroid and a niobium plate. This bridge ensures good heat conduction between the sample and the low-temperature specimen stage. When isolated toroidal samples were placed on a carbon film, as shown in Fig.6.15, they could not be cooled below T_c of the covering superconductor.

In preparing the samples, special attention was paid to attaining perfect contact between the two niobium layers, because even a slight oxidation layer between them would break down the superconducting contact and allow the magnetic field to leak out. The niobium oxide produced on top of the lower niobium layer in the lithography processes had therefore to be removed by ion sputtering before the upper niobium layer was laid down. In another experiment, a maximum current density of 40 mA/μm^2 was confirmed to pass through the contact between the two layers at 4.2 K. This current was much larger than the 10 mA/μm^2 persistent current calculated to be necessary for quantizing the magnetic flux. Except for a slight flux change due to changes in sample temperature, the magnetic flux of this sample cannot change very much. Various toroids, having different flux values, were therefore connected to the niobium plate (Fig.6.16).

b) Experimental Results

Experiments were carried out in a manner similar to that described in Sect.6.6.1. An electron hologram of a sample was first formed by using a 150 kV field-emission electron microscope, and then the relative phase shift

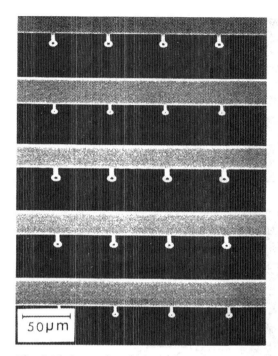

Fig. 6.16. Array of toroidal samples

for the sample was optically reconstructed using a HeNe laser. A preliminary experiment was carried out to test whether there would be any observable interaction between an electron beam and a simple toroidal superconductor containing no magnet. This experiment was to confirm that the phase of an electron beam passing near a superconductor is not influenced by it, even though the Meissner effect prevents the magnetic field accompanying the electron beam from passing through the superconductor.

This preliminary experiment measured the phase difference between electron beams passing inside and outside of the 300 nm thick niobium toroid which was completely covered by a layer of 50 to 200 nm thick copper. When the temperature of the toroid was varied above and below its critical temperature (9.2K), the twofold phase-amplified interference micrographs showed no phase difference in the normal or the superconducting state (Fig.6.17). Another experiment confirmed that niobium samples actually became superconducting at 4.5 K in this experiment: an external magnetic field was applied to a toroidal sample and the flux trapped in the sample hole at 4.5 K was observed by electron holography. These preliminary experiments confirmed that the toroidal superconductor itself would not influence an incident electron beam, that is, it would not produce a re-

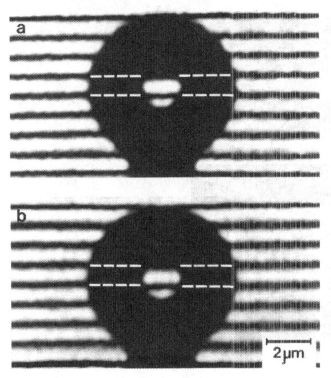

Fig. 6.17a,b. Experimental results showing no interaction between an electron beam and superconducting toroid without a magnet inside: (a) Normal state at 15 K, and (b) superconducting state at 4.5 K

lative phase shift between a beam passing through the hole and a beam passing outside the toroid.

Fabricated toroidal samples having no leakage flux were then selected by using electron-holographic interferometry, and the magnetic flux leaking from the samples was measured. This experiment confirmed that there was no interaction, to within the precision of a $2\pi/10$ phase shift, between the electron beam and the superconducting toroid. The samples were not always free of leakage flux (Fig.6.18). Leakage was found to be appreciable for samples with large diameters (larger than $10\,\mu$m) or with small aspect ratios. After the sample geometry was redesigned and the magnetic characteristics of the permalloy film were improved, approximately 60% of the samples were confirmed to be free from leakage to within a precision of $(1/20)\times(h/e)$.

A leakage-free toroid was selected and cooled in a low-temperature stage to 4.5 K. Then, an electron beam illuminated the sample and the relative phase shift between the two beams could be measured. Only two kinds of interferograms were obtained: the relative phase shift was either 0

Fig. 6.18. Interference micrograph of the leakage flux from a toroidal sample (phase amplification: ×2)

or π (Fig.6.19). Photograph (b), indicating a relative phase shift of π, is evidence for the AB effect. Under the experimental conditions, the magnetic field was confined within the superconductor and the field was shielded from the electron beam by the copper and niobium covering.

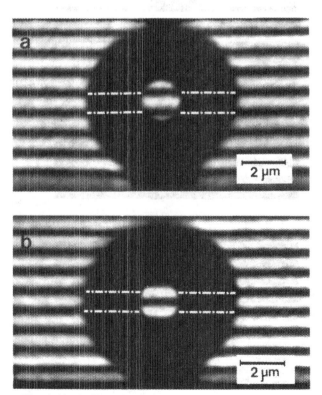

Fig. 6.19a,b. Experimental results showing the Aharonov-Bohm effect: (**a**) Phase shift of 0, and (**b**) phase shift of π

The quantization of the phase shift provides key evidence that the magnetic field was completely shielded by the covering superconductor, and this quantization was clarified by temperature-dependence experiments. The sample illustrated in Fig.6.20 is the same one shown in Fig.6.19b. When this sample was cooled from T = 300 K, the phase shift gradually increased from 0.3π at T = 300 K (Fig.6.20a), becoming $0.8\ \pi$ at T = 15 K (Fig.6.20b). When the temperature was further reduced to below T_c, the phase shift abruptly changed from $0.8\ \pi$ to π (Fig.6.20c). This behavior

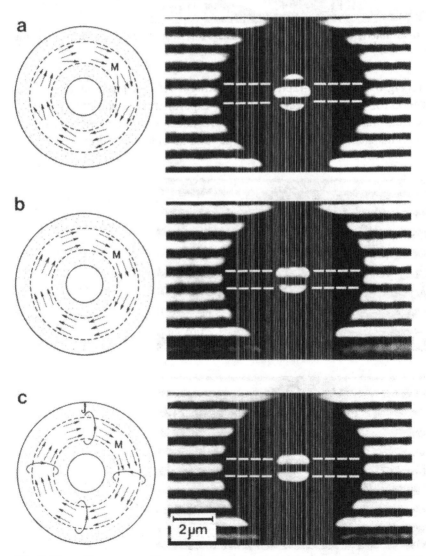

Fig. 6.20a-c. Temperature dependence of relative phase shifts: (**a**) Phase shift of 0.3π at 300 K, (**b**) phase shift of 0.8π at 15 K, and (**c**) phase shift of π at 4.5 K

can be interpreted as follows: The enclosed magnetic flux determines the phase shift above T_c, and the gradually increasing phase shift as the sample temperature changes from 300 to 9.2 K arises from the magnetization in the permalloy increasing by 5% because of decreasing thermal fluctuations of the spins. When the temperature falls below T_c, a supercurrent flows in the inner surface layer of the hollow superconducting torus so that the total magnetic flux may be equal to an integral multiple of $h/(2e)$. The phase shift is therefore either 0 or π depending on whether the number of fluxons is even or odd.

No quantization was observed in samples having even a slightly oxidized layer between the two niobium layers. The occurrence of flux quantization can therefore be taken as assurance that the niobium layer actually becomes superconducting, that the superconductor completely surrounds the magnetic flux, and that the Meissner effect prevents any flux from leaking outside. In Fig.6.20b the phase shift just above T_c was 0.8π and was closer to π than to 0, resulting in a jump to π.

In this way, the existence of the AB effect was confirmed without any doubt.

7. Electron-Holographic Interferometry

When holography is employed, the phase distribution of an electron beam transmitted through or reflected from a sample can be displayed as an interference micrograph. Although an interference micrograph can also be obtained without recourse to holography if we use an electron microscope equipped with an electron biprism (Sect. 3.2), electron holography allows contour maps to be observed and the phase to be measured with a precision as high as $2\pi/100$. The development of electron-holographic interferometry allows us to see objects that were not visible when using conventional electron microscopes in which only the intensity of an electron beam is observed.

7.1 Thickness Measurements

7.1.1 Principle of the Measurement

An electron beam transmitted through a specimen is phase-modulated, because inside of the specimen the beam is accelerated by the **inner potential** V_0, and consequently its wavelength becomes shorter. When the specimen is uniform, the phase shift is proportional to the specimen thickness. The phase difference $\Delta S/\hbar$ between an electron beam passing through a uniform nonmagnetic specimen and the beam passing through in vacuum can be derived from (6.6) as

$$\frac{\Delta S}{\hbar} = \frac{\sqrt{2me}}{\hbar}\int(\sqrt{V+V_0}-\sqrt{V})ds \approx \frac{\sqrt{2meV}}{2\hbar}\frac{V_0}{V}t, \tag{7.1}$$

where t is specimen thickness. In other words, the refractive index n of the specimen is given by $n = 1 + \frac{1}{2}V_0/V$.

When $V = 100\,\text{kV}$ and $V_0 = 20\,\text{V}$, the value of n−1 is 10^{-4}. The 40 nm change in specimen thickness corresponding to a phase shift of 2π shows that the sensitivity of thickness measurements is very low compared to the extremely short electron wavelength of 40 pm. This low sensitivity results from the small value of n−1, so phase-amplification techniques are indispensable when thickness distributions are to be measured in atomic dimensions.

7.1.2 Examples of Thickness Measurement

An example of a thickness measurement is displayed in Fig. 7.1. The reconstructed image (a) is equivalent to the electron micrograph, which reveals the intensity distribution of an electron beam transmitted through magnesium-oxide smoke particles. The interference micrograph (b) employs contour lines of 50 nm steps. The specimen thickness can be measured at any point in the electron micrograph. If the phase-amplification technique (Sect. 5.2.2) is utilized, a twofold phase-amplified contour map (c) is obtained and a more detailed thickness distribution can be seen.

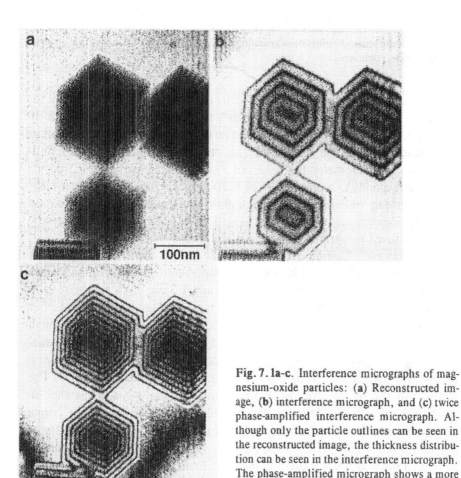

Fig. 7.1a-c. Interference micrographs of magnesium-oxide particles: (a) Reconstructed image, (b) interference micrograph, and (c) twice phase-amplified interference micrograph. Although only the particle outlines can be seen in the reconstructed image, the thickness distribution can be seen in the interference micrograph. The phase-amplified micrograph shows a more detailed thickness distribution

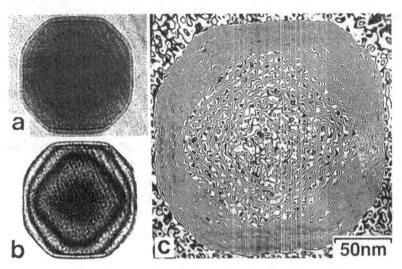

Fig. 7.2a-c. Interference micrographs of a fine beryllium particle [7.2]: (**a**) Reconstructed image, (**b**) interference micrograph, and (**c**) 32-times phase-amplified interference micrograph. When the phase distribution was amplified 32 times, a map of thickness contours in 1.3 nm steps was obtained

The minimum phase difference detectable in the interference micrograph (c) is about $\lambda/4$, which corresponds to a thickness of about 6 nm. If the sensitivity is increased by more than an order of magnitude, observation and measurement on the atomic scale become possible. In light optics, multiple-beam interferometry has achieved a surprising sensitivity of 230 pm, which corresponds to a phase shift of $\lambda/2000$ [7.1]. An attempt to use electron holography to detect such a small phase shift was carried out not only by using overlapping twin images (Fig. 5.7) but also by reforming new phase-amplified holograms in the spatial filtering system (Fig. 5.8).

An example of a fine beryllium particle [7.2] is depicted in Fig. 7.2. A conventional interference micrograph (b) represents thickness contour lines in 44 nm steps, whereas contour lines of only 1.3 nm steps can be seen in the 32-times amplified interference micrograph (c). One may doubt whether these fringes represent the thickness distribution precisely or are only rough interpolations. This problem was examined through the observation of **surface steps** by *Tonomura* et al. [7.3]. Photographic evidence of interferograms of a molybdenite film about 5 nm thick is exhibited in Fig. 7.3. The phase shift can be observed as displacements of equally spaced parallel interference fringes in a 24-times phase-amplified interferogram. The fringe steps along the lines designated A, B and C correspond to phase shifts of 0.025λ, 0.075λ and 0.125λ, respectively. A phase shift of 0.025λ corresponds to the 620 pm thickness change due to an atomic step, which is one-half of the c-axis lattice spacing. The points A, B and C, therefore, indicate

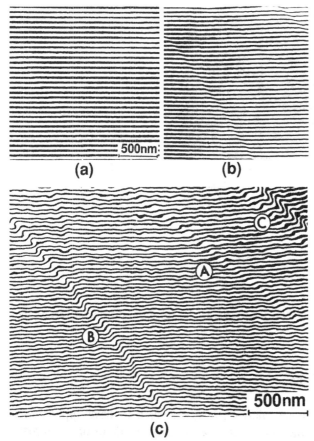

Fig. 7.3a-c. Interferograms of molybdenite thin film: (**a**) Conventional interferogram, (**b**) four-times phase-amplified interferogram, and (**c**) 24-times phase-amplified interferogram. Although hardly any fringe displacements are observed in the conventional interferogram, surface atomic steps can be observed in the 24-times amplified interferogram. Step A corresponds to a phase shift of 0.025λ and to a thickness change equal to the monoatomic step height (620pm)

steps of one, three, and five atomic layers, respectively, on cleaved surfaces.

A **carbon nanotube** discovered by *Iijima* [7.4] was observed using the phase-shifting method in an electron microscope. The phase image reconstructed from 100 different holograms obtained by tilting an incident electron beam onto the specimen and the phase profile across the tube are shown in Fig. 7.4. Although the phase shift due to a single carbon layer is as small as $1/200$ of 2π, the phase profile indicates the tube structure. Carbon nanotubes were also investigated using holography elsewhere [7.6-7].

Voids in Si samples [7.8] or in Pd particles [7.9] were studied by holographic interference microscopy.

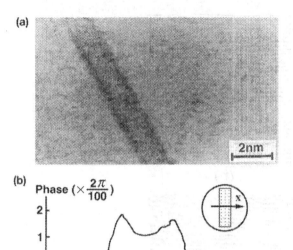

Fig. 7.4a,b. Carbon nanotube: (a) Phase image, and (b) phase distribution across the carbon nanotube [7.5]

7.1.3 Other Applications

Phase-amplified interference microscopy is expected to have applications other than the thickness measurement of a specimen made of uniform material such as phase-contrast observation of biological specimens [7.10]. Before the electron microscopic observation, staining or evaportation with heavy atoms is usually required for **biological specimens**. They are phase objects to an illuminating electron beam, and contrast is hard to observe. However, if biological specimens are to be be detected on atomic scales, attached heavy atoms are obstacles. It is desirable that the contrast should arise from the specimens themselves.

Because a single carbon atom shifts the phase of a 100 kV electron beam by $\lambda/70$ [7.11], phase contrast can be used for the observation of unstained biological specimens in electron microscopy, just as it is used in optical microscopy, allowing the observation of unstained biological specimens. Figure 7.5 displays a 24-times phase-amplified interference micrographs of an unstained ferritine molecule [7.10]. Although the core of the ferritine molecule can be observed by means of conventional electron microscopy, these interference micrographs also allow a quantitative observation of the surrounding protein.

Fig. 7.5a,b. Interference micrographs of an unstained ferritine molecule: (**a**) 24-times phase-amplified contour map, and (**b**) 24-times phase-amplified interferogram. The surrounding protein part of the ferritine molecule, which is hardly seen in conventional electron microscopy, can be seen as well as the core

7.2 Surface Topography

In the transmission mode of electron holography, surface topography can only be investigated by measuring thickness. Although the intensity of an electron beam reflected from a specimen surface is generally too weak to form a hologram, *Osakabe* et al. [7.12] have demonstrated the feasibility of electron holography with a Bragg-reflected beam from a single-crystal surface. In this reflection mode, surface topography can be measured with a high degree of sensitivity because surface-height differences are directly measured as geometrical path differences in units of the extremely short electron wavelength.

An example of a surface topography measurement is presented in Fig. 7.6. The specimen is single-crystal GaAs with an atomically flat surface that has a single **screw dislocation**. One fringe displacement in the interferogram corresponds to a height difference of only 50 pm. The 200 pm high monatomic step indicated by the arrow is terminated by a dislocation core. One can see at a glance how the 200 pm difference in height is relaxed in a region surrounding the core. Inside the crystal this relaxation is supposed to be just like a spiral staircase around the dislocation line. However, this micrograph reveals that at the surface the slope of the staircase is not uniform but is steep only in one direction. The relaxation extends even to a region a few μm away. In this mode of holography, surface topography can be measured to a precision of 10 pm even in an unamplified interference micrograph.

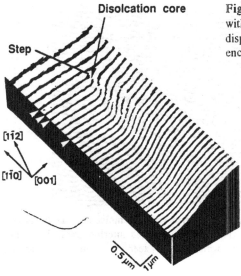

Fig. 7.6. Interferogram of a GaAs surface with a single screw dislocation. One fringe displacement corresponds to a height difference of 50 pm

7.3 Electric Field Distribution

When electric fields are observed by electron interference microscopy, contour fringes indicate projected **equipotential lines**. This can be easily understood from the equation expressing the phase shift $\Delta S/\hbar$ due to a microscopic distribution of electrostatic potential ΔV:

$$\frac{\Delta S}{\hbar} = \frac{1}{2\hbar}\sqrt{2me/V} \int \Delta V \, ds \,. \tag{7.2}$$

If the potential has the same value along the electron-beam direction, the phase shift is proportional to ΔV. The phase contour lines therefore indicate equipotential lines. *Merli* et al. [7.13] detected a phase shift due to the electric potentials near a pn junction, and *Kulyupin* et al. [7.14] used electron interferometry to measure the electric-potential distribution near the apex of a field-emission tip. The microscopic distribution of an electric field near a pn junction (Fig. 7.7) was holographically observed by *Frabboni* et al. [7.15].

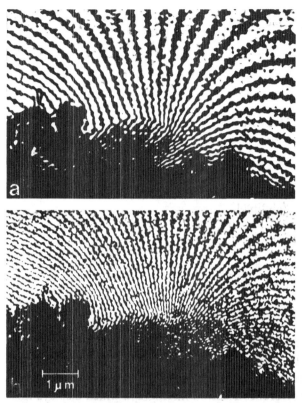

Fig. 7.7a,b. Equipotential lines near p-n junction observed by electron-holographic interferometry [7.5]: (**a**) Reverse bias of 4 V, and (**b**) reverse bias of 8 V. Projected equipotential lines are directly seen as contour fringes

7.4 Domain Structures in Ferromagnetic Thin Films

7.4.1 Measurement Principles

The microscopic distribution of magnetic fields can be seen by using electron-holographic interferometry, and an interference micrograph can be interpreted in a straightforward way [7.16-20]. The observation principle is depicted in Fig. 7.8. When an electron beam is incident into a uniform magnetic field, the beam is deflected by the Lorentz force acting perpendicularly to the magnetic field. When electrons are viewed as waves, the introduction of a wave front perpendicular to the electron trajectory will suffice. Incident parallel electrons are represented by a plane wave, and electrons transmitted through the magnetic field by a tilted plane wave with the left side up. In other words, the wave front revolves around the magnetic line of force. If one imagines a contour map of this wave front, the contour lines

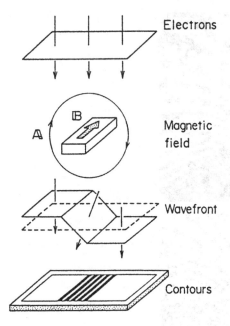

Fig. 7.8. Principle underlying the observation of magnetic flux

would follow the magnetic lines of force. This is because the height of the wave front differs on both sides of the magnetic line of force but is the same along it. Thus we reach a very simple conclusion: when a magnetic field is observed in an interference electron micrograph which displays the contour map of the transmitted wave front, the contour fringes indicate the magnetic lines of force.

The aforementioned contour fringes are also quantitative. A simple calculation yields that a constant amount of minute magnetic flux (h/e) flows between adjacent contour fringes. A superconductive flux meter, SQUID [7.21], can measure the flux in units of h/(2e) by using Cooper pair interference. The electron interference micrograph is formed due to the interference not of Cooper pairs, but of electrons. In electron-holographic interferometry, the flux unit becomes h/e because the electric charge is −e instead of −2e. The principle is the same for electron interferometry and SQUID.

This can be explained more exactly using (6.7) which indicates the existence of the phase shift $\Delta S/\hbar$, because of the magnetic field, between two electron beams starting from one point and ending at another point:

$$\frac{\Delta S}{\hbar} = -\frac{e}{\hbar}\int \mathbf{B}\cdot d\mathbf{S}, \tag{7.3}$$

where the integral is performed over the surface enclosed by the two electron trajectories. The integral is equal to the magnetic flux passing through the surface. If two electron beams pass through two points along a single magnetic line, then the phase shift vanishes because the magnetic flux enclosed by the two beams is zero. When the two beams enclose a flux of h/e, the phase shift becomes 2π. An interference micrograph of magnetic fields thus provides intuitive, microscopic and quantitative information about the fields.

Cohen [7.22] noted for the first time that holographic techniques could give direct information about magnetic domain structures. *Tonomura* [7.23], and *Pozzi* and *Missiroli* [7.24] subsequently demonstrated that information about the domain structure actually is reflected in the electron phase distribution. In 1980, the magnetic lines of force inside ferromagnetic particles were first observed as contour fringes in interference micrographs by *Tonomura* et al. [7.18]. This method has the following advantages over conventional Lorentz microscopy which requires defocussing:

- Intuitive interpretation is possible because magnetic lines of force appear as contour fringes in an electron micrograph.
- Magnetic structures can be related to the fine structures of the specimen because there is no need to defocus an electron microscopic image.
- Quantitative measurment is possible because a constant flux of h/e ($=4 \cdot 10^{-15}$ Wb) flows between two adjacent contour lines.

7.4.2 Magnetic Domain Walls in Thin Films

Let us first consider applications to the observation of magnetic domain walls in thin films [7.25]. The simplest example here is a Néel wall in a permalloy thin film (Fig. 7.9). Photograph (a) is a Lorentz micrograph in which the magnetic domain structures can be observed with a high spatial resolution. In this photograph, we can see a single black line that indicates the magnetic domain wall on both sides of which the magnetization is in different direction. This interpretation is easily understood from the principle behind Lorentz microscopy (Fig. 7.10). Incident electrons are deflected by the Lorentz force due to magnetization inside of a specimen. Since magnetization directions are opposite in the two domains, the deflections are also in the opposite directions. Therefore, if the electron intensity is observed in the lower plane, a deficient line is seen along the domain wall.

The Lorentz micrograph in Fig. 7.9a is, so to speak, a defocused electron micrograph, and the image details are therefore blurred. In addition, one cannot see how magnetic lines of force flow in two domains and rotate at the domain wall. The interference micrograph (b) gives a direct answer: the magnetization distribution is seen at a glance as magnetic lines of force.

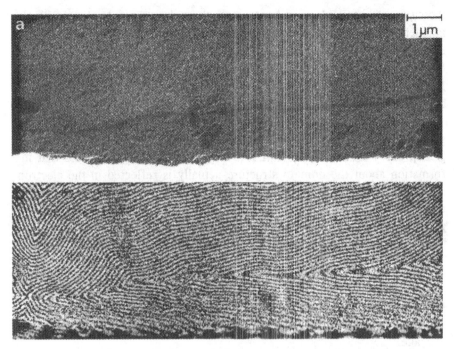

Fig. 7.9a,b. Interference micrograph of a 30 nm thick permalloy film: (**a**) Lorentz micrograph, and (**b**) interference micrograph. Only a dark line is seen along a Néel wall in (**a**), while magnetization flow in two domains and its rotation at the wall can be seen at a glance in (**b**)

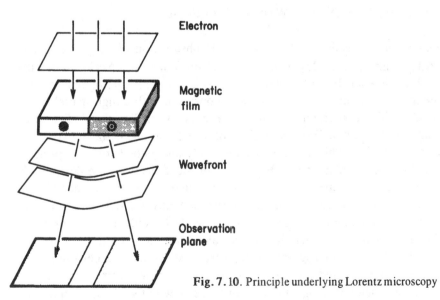

Fig. 7.10. Principle underlying Lorentz microscopy

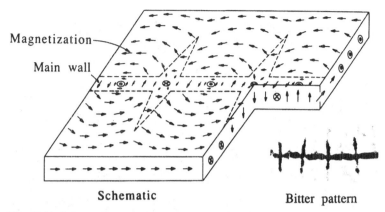

Fig. 7.11. Spin configuration model for cross-tie wall

The magnetic flux flowing to the left in the upper part of the figure approaches the domain wall and, all at once, changes its course.

An interesting domain wall structure can be seen in a permalloy film thicker than the one in which the Néel wall was visible. The spin configuration of this cross-tie wall (Fig.7.11) was first predicted by *Huber* et al. [7.26] from the Bitter figure. This cross-tie wall is a 180° wall and magnetization flows along a straight line at a distance from it. Magnetization becomes more wavy near the main wall, and it finally rotates in circles. Such a structure can be seen with a brief glance at the interference micrograph exhibited in Fig.7.12, where the contour lines are in accordance with the prediction by *Huber* et al. (Fig.7.11). The simple reason for such a complex structure is as follows. For the Néel wall, the magnetization in the wall region is all in the same direction. The wall region can be compared to a row of tiny bar magnets whose magnetization directions are the same and perpendicular to the wall. When a film becomes thicker, and the bar magnets consequently become stronger, such an arrangement becomes unstable. If we take into consideration the magnetostatic energy, we can see that we have a more stable state when every other magnet is reversed. The resultant wall is a cross-tie wall.

Fig. 7.12. Inteference micrograph of a 50 nm thick permalloy film. Magnetic lines of force near a cross-tie wall can be seen directly

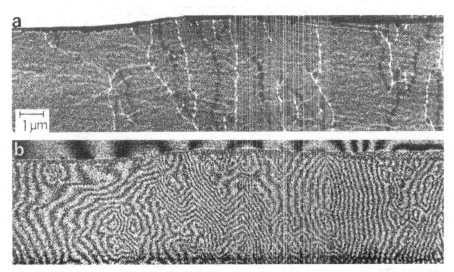

Fig. 7.13a,b. Interference micrograph of a nickel film: (**a**) Lorentz micrograph, and (**b**) interference micrograph. Circular magnetization can be clearly seen in the interference micrograph barely visible in the Lorentz micrograph

The polycrystalline nickel film shown in Fig. 7.13 has a more complex domain wall structure, and it is therefore barely possible to see fine details of the magnetization in the Lorentz micrograph (a). It is much easier to interpret the magnetization distribution in the film by observing the interference micrograph (b), which exhibits several small domains.

7.5 Domain Structures in Fine Ferromagnetic Particles

Figure 7.14 illustrates a fine, plate-shaped cobalt particle. The reconstructed image (a), which is equivalent to the electron micrograph, has no contrast inside the particle image. This is because the sample thickness is uniform and the magnetization has no influence on the intensity of the transmitted electron beam. Information about the magnetization distribution is contained in the phase distribution. In fact, from the twofold phase-amplified interference micrograph (b) it is possible to see how magnetic lines of force rotate in such a fine particle. The contour map (b) itself does not allow one to decide whether the magnetization direction is clockwise or counter-clockwise, since these two possibilities correspond to whether the wave front is protruded like a mountain or hollowed like a valley. However, this can be decided from the interferogram (c), which is obtained by slightly tilting two interfering beams in the optical reconstruction stage. The wave

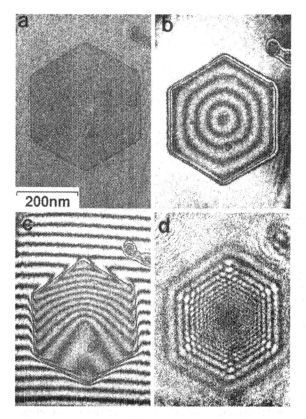

Fig. 7.14a-d. Inteference micrographs of a hexagonal cobalt particle: (**a**) Reconstructed image, (**b**) two-times phase-amplified contour map, (**c**) two-times phase-amplified in interferogram, and (**d**) Lorentz micrograph. No contrast can be seen in the reconstructed image, whereas in-plane magnetic lines of force are displayed as contour fringes in the contour map. The direction of magnetic lines can be determined from the interferogram to be clockwise. The Lorentz micrograph can be obtained optically from the hologram, from which it is difficult to determine the magnetic domain structure

front of the transmitted electron beam is first retarded at the particle edge because of the thickness effect. Then it is advanced inside the particle because of the magnetic effect. This means that the magnetization is clockwise.

It has been difficult to experimentally determine the magnetization distribution in such a fine particle. The magnetic structure is difficult to identify even when observed by Lorentz microscopy, which provides magneti domain-structure information with the highest spatial resolution currently available [7.27]. This is a result of the large defocusing needed for the observation of domain structures resulting in the magnetic contrast overlapping the diffraction pattern of the particle. A Lorentz micrograph of the particle is depicted in Fig.7.14d. This micrograph was optically recon-

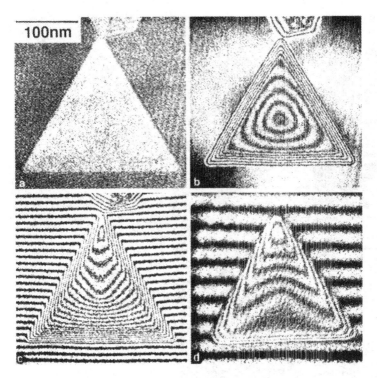

Fig. 7.15a-d. Interference micrographs of a fine cobalt particle: (**a**) Reconstructed image, (**b**) contour map (phase amplification: ×2), (**c**) interferogram (phase amplification: ×2), and (**d**) interferogram of another sample having the opposite rotation direction of magnetization

structed by merely defocusing the reconstructed image from the same electron hologram, which is possible because a hologram contains all the information of the scattered electron wave from an object. Because the outer shape of the particle is completely blurred, as can be seen in the Lorentz micrograph (d), it is not easy to predict the magnetization distribution.

Figure 7.15 shows interference micrographs of a triangular cobalt particle [7.18]. Since this particle is a triangular pyramid truncated parallel to the base plane, two kinds of interference fringes can be seen in the interference micrograph (b). Contour fringes parallel to the three edges show that the thickness increases up to 55 nm from the edges. The inner contour fringes indicate magnetic lines of force, since the thickness is uniform there. The magnetization direction is determined from interferogram (c) and is clockwise. Of course, a similar particle with a counter-clockwise magnetization could also be found, as shown in interferogram (d).

7.6 Magnetic Devices

Electron-holographic interference microscopy can help in evaluating practical magnetic devices, such as those used for high-density magnetic recording [7.28, 29]. Magnetic recording plays an important role in the storage of information for a variety of applications ranging from tape recorders to computer memories. With recent increases in the amount of information stored, recording densities keep rising and the recorded bit length keeps decreasing. Detailed observations of recorded magnetization patterns are required, since the recording density is approaching some fundamental limitations.

An interference micrograph of a recorded magnetic tape is presented in Fig. 7.16. A moving magnetic head was used to record on a cobalt film. The film is observed from above, and the upper part in the micrograph (b) is the film; the lower part is free space. Arrows in the micrograph indicate the directions of the recorded magnetization, and the magnetic lines of force recorded can be seen as contour fringes along arrows indicated in the schematic diagram (a). Two streams of magnetization collide head-on and produce vortices similar to those of water streams, meandering as they approach the film edge when they finally leak outside into space.

To achieve a higher-density recording, the width of the transition region has to be made small. This width was found to depend on the magnetic

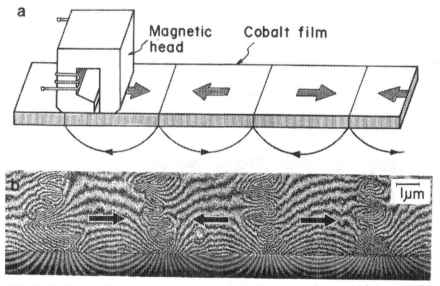

Fig. 7.16a,b. Recorded magnetization pattern in a cobalt magnetic tape: (a) Schematic diagram of the recording method, and (b) two-times phase-amplified interference micrograph

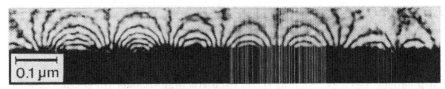

Fig. 7.17. Ten-times phase-amplified interference micrograph of a recorded cobalt tape (bit length: 0.15 µm). Only leakage fluxes can be seen

characteristics of the film materials and on the recording method. By using interference microscopy and adjusting experimental conditions, a bit length of 0.15 µm could be demonstrated. The resultant interference micrograph being phase-amplified 10 times is depicted in Fig. 7.17, which shows only magnetic lines of force leaking outside the recorded film.

The recording method so far described is called **in-plane magnetic recording**, since the magnetization direction is in the plane of the tape. In the **perpendicular magnetic recording** developed for high density by *Iwasaki* and *Nakamura* [7.30], on the other hand, the direction of magnetization is perpendicular to the tape plane. In this case, the recorded magnetization cannot be observed by using an electron beam incident perpendicular to the tape. The tape has to be sliced, as shown in Fig. 7.18a, and observed by an

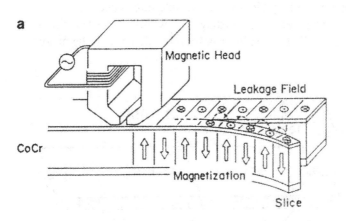

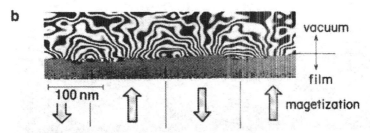

Fig. 7.18a,b. Interference micrograph of a perpendicularly recorded magnetic tape: (**a**) Recording method, and (**b**) interference micrograph (phase amplification: ×70)

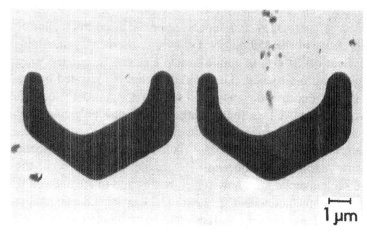

Fig. 7.19. Electron micrograph of a propagation circuit for magnetic bubble memory. An array of tiny horseshoe magnets made of permalloy is placed on a thin carbon film

electron beam incident perpendicular to this section (parallel to the original tape surface). The resultant magnetic lines of force are displayed in Fig. 7.18b. For this type of magnetic recording, a bit length of 70 nm has been confirmed.

Another application is to a "bubble" memory device [7.19], in which the magnetic fields from tiny horseshoe magnets (Fig. 7.19) propagate magnetic bubble domains situated below the magnet layer as a result of the rotation of an applied in-plane magnetic field. The magnetic field distribution is important for accurate and smooth propagation of the magnetic bubbles, so the magnetic lines of force have been observed using electron holography.

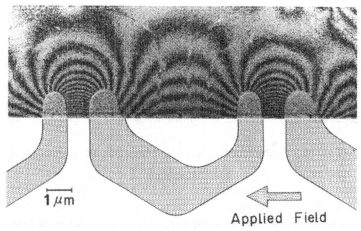

Fig. 7.20. Interference micrograph of magnetic fields from tiny horseshoe magnets

This propagation circuit whose interference micrograph is shown in Fig. 7.20 was made of permalloy and supported on a 10-nm thick film of carbon. The periodicity of the 350 nm thick chevron pattern is 8 μm. An in-plane magnetic field of 20 G was applied when electron holograms were formed in the electron microscope. The field direction is indicated by the arrow in the picture. Because the magnetic field was applied to the reference electron beam as well as the propagation circuit, only magnetic fields from the magnetized circuit are seen in the micrograph.

The contour lines in Fig. 7.20 clearly illustrate how some magnetic lines of force originating from one end of a magnet are directed to the other end of the same magnet, and how some are directed to the end of the adjacent magnet. The magnetic-field strength can be estimated everywhere in the micrograph because a magnetic flux of $h/(2e)$ is contained between two contour lines in this twofold phase-amplified interference micrograph. For example, the magnetic-field strength between two magnets is 1100 G when the interaction length between electrons and the magnetic field is assumed to be 350 nm.

7.7 Domain Structures in Three-Dimensional Particles

Although electron-holographic interference microscopy provides direct and quantitative information about the magnetic domain structure, only specimens whose thicknesses were uniform could be investigated because the phase shift is, in general, due not only to the magnetic fields but also to the change in specimen thickness. The magnetic domain structure of a three-dimensional fine particle, for example, cannot be determined from its interference micrograph, as depicted in Fig. 7.21. The specimen here, a cobalt pentagonal decahedron consisting of five regular tetrahedrons having a common side, is not uniform in thickness. As a result, there is no direct way to interpret the interference micrographs. This problem can be expressed more generally: a single hologram cannot distinguish between the effects of electric and magnetic fields (or thickness and magnetic field, respectively).

Tonomura et al. [7.31] therefore developed a holographic technique to independently observe the electric and magnetic contributions in interference electron microscopy. This technique employs the different behavior of electric and magnetic fields for the time-reversal operation. That is, two holograms are formed, one in the standard specimen position and one with the face of the specimen turned down. In these holograms, the electric contributions are the same but the magnetic contributions have opposite signs.

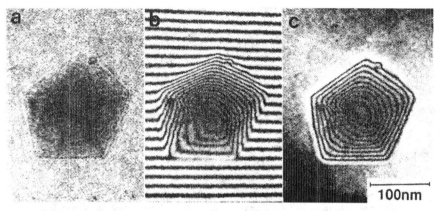

Fig. 7.21a-c. Interference micrographs of a decahedral cobalt particle: (**a**) Reconstructed image, (**b**) interferogram (phase amplification: ×2), and (**c**) contour map (phase amplification: ×2)

A holographic technique then displays the electric and magnetic images separately.

This can easily be understood from the Equation (6.6) of the phase shift $S/\hbar$

$$\frac{S}{\hbar} = \frac{1}{\hbar}\int(m\mathbf{v} - e\mathbf{A})\cdot d\mathbf{s}, \tag{7.4}$$

where the line integral is performed along the electron path. If an electron beam is incident on the specimen from the opposite direction (t → −t, **v** → −**v**, **s** → −**s**), then the electric contribution to the phase shift [the first term in (7.4)] remains the same but the magnetic contribution reverses the sign of the phase shift. Such an effect can be observed when the specimen is viewed from two opposite directions as interference micrographs (Fig. 7.22). The phase distributions differ between top and bottom views, and this indicates the presence of a magnetic field. It is interesting to note that an object may appear quite differently, depending on the viewing direction.

If we consider the phase distribution in the micrograph (Fig. 7.22b) to consist of an electric phase plus a magnetic phase, then the phase distribution in micrograph (Fig. 7.22a) represents of the electric phase minus the magnetic phase. If these optically reconstructed images are made to overlap in order to display the phase difference between them, the resultant phase distribution becomes two times the magnetic phase. In this way, a purely magnetic image can be obtained. To create the thickness image, we can make full use of the conjugate image, which has a phase distribution with the sign opposite to the original sign. The pure thickness image can be ob-

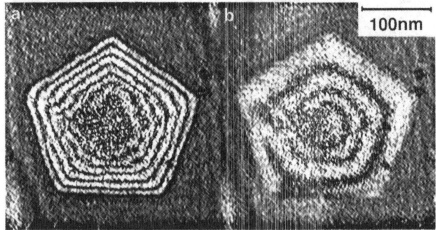

Fig. 7.22a,b. Two interference micrographs of a decahedral cobalt particle observed from opposite directions (phase amplification: ×2). (a) Top view and (b) bottom view. The appearance of the same sample depends on the viewing direction

tained by overlapping the reconstructed image of the top view and the conjugate image of the bottom view.

The results are illustrated in Fig. 7.23. Micrographs (a) and (b) represent the thickness contours and in-plane magnetic lines of force, respectively. Although many choices are potentially available for the domain structure of a three-dimensional particle, magnetization can be determined to rotate along the common side of the five regular tetrahedrons forming the particle, as illustrated in Fig. 7.24.

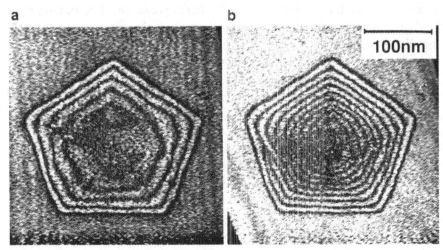

Fig. 7.23a,b. Interference micrographs of a cobalt particle (phase amplification: ×2): (a) Thickness contour map, and (b) magnetic lines of force

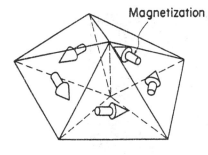

Fig. 7.24. Schematic diagram of magnetic domain structure in a cobalt particle

7.8 Three-Dimensional Image

An interference micrograph provides information about an object or more exactly electromagnetic potentials, but only that projected along the direction of an incident electron beam. There are cases where three-dimendional structures cannot be determined solely by interference microscopy. The CT technique introduced in Sect.5.2.3 can then be utilized to reconstruct such structures.

7.8.1 Electric Potentials

The phase shift of an electron beam transmitted through an electric object $\Delta V(x, y, z)$ relative to that of an electron beam passing through in a vacuum is given by

$$\frac{\Delta S}{\hbar} = \frac{(2me/V)^{1/2}}{2\hbar} \int \Delta V(x, y, z) \, ds \; . \tag{7.5}$$

This means that the phase distribution is proportional to the projection of the electric potential distribution along the beam direction.

Lai et al. formed 24 holograms of latex spheres taken from different directions, by tilting the specimen around an axis every 5° from −60° to + 60°. Then, they employed an iterative CT algorithm to reconstruct the three-dimensional distribution of the electric potential [7.32]. The electric potential distribution in any cross section can be displayed. Two outside views of the reconstructed image from different directions are depicted in Fig.7.25.

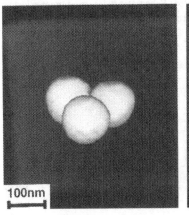
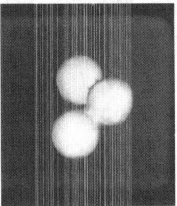

Fig. 7.25. Two views of latex spheres

7.8.2 Magnetic Fields

Due to the vectorial nature of a magnetic field, the phase shift of an electron beam transmitted through the magnetic field **B** relative to that of the beam passing through the origin (x=0, y=0, z=0) is not so simply related as (7.2) but as

$$\frac{\Delta S(x,y)}{\hbar} = -\frac{e}{\hbar}\int \mathbf{B}(x,y,z)\cdot d\mathbf{S} ,\qquad(7.6)$$

where the surface integral is carried out over the surface enclosed by the two beams.

However, it was clarified by *Lai* et al. [7.33] that **B** can be determined from many electron holograms obtained when the magnetic object is tilted not only around a single axis but around two axes using the extended CT algorithm as follows: When (7.6) is applied to the case illustrated in Fig. 7.26 where the plane d**S** is in the xz plane, (7.6) can be written as

$$\frac{\Delta S(x,y)}{\hbar} = -\frac{e}{\hbar}\int_0^x d\xi \int B_y(\xi,y,z)\,dz .\qquad(7.7)$$

It can be inferred from this equation that the projection of B_y along the z-axis is given by $\partial \Delta S(x,y)/\partial x$ except for a constant e, which can be derived by differentiating the measured phase distribution $\Delta S(x,y)$ along the x-

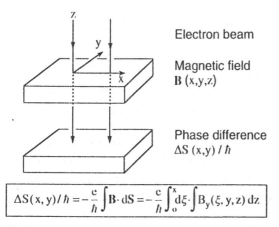

Fig. 7.26. Relation between the phase distribution $\Delta S(x, y)/\hbar$ and magnetic field $\mathbf{B}(x, y, z)$

$$\Delta S(x,y)/\hbar = -\frac{e}{\hbar}\int \mathbf{B}\cdot d\mathbf{S} = -\frac{e}{\hbar}\int_0^x d\xi \cdot \int B_y(\xi, y, z)\, dz$$

direction. Similarly, the projection of B_x is given by $\partial \Delta S(x,y)/\partial y$. If the magnetic object is tilted around the x-axis, B_x at each pont remains unchanged by tilting. Therefore, $B_x(x,y,z)$ can be calculated from the measured phase distribution $\Delta S(x,y)$ with the CT algorithm. $B_y(x,y,z)$ can also be reconstructed from the data obtained by tilting the object around the y-axis. The component $B_z(x,y,z)$ can automatically be derived from Maxwell's equations.

An example of the result is displayed in Fig. 7.27 where the magnetic field leaking out from the north pole of a barium-ferrite particle is illustrated.

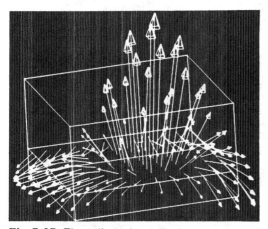

Fig. 7.27. Three-dimensional distribution of the magnetic field leaking out from the north pole of a barium-ferrite particle

7.9 Dynamic Observation of Domain Structures

Real-time observation techniques introduced in Sect. 5.2.4 have made it possible not only to observe specimens without a time delay just as in observing electron-microscopic images but also to observe the dynamic phenomena in specimens.

In the technique already exhibited in Fig. 5.12, electron holograms are detected with a TV camera and recorded in a liquid-crystal panel, which acts as a real-time phase hologram. Illuminating the panel with a laser beam produces an interference micrograph. Dynamic events can be seen on the TV system.

An example of video scenes is displayed in Fig. 7.28. Magnetic lines of force in a permalloy thin film can be observed in (a) where no magnetic field is applied to it. Three magnetic domains can be seen, inside each of which the magnetic lines of force are uniform and in the same direction. The two domain walls meet at the point P. When a magnetic field was applied towards left (the direction of the magnetic field is indicated by the arrow in Fig. 7.28b) and increased gradually up to 36 G. The central domain expanded its area as in (b) since the direction of the applied magnetic field H was nearly the same as that of the magnetic lines of force in the central

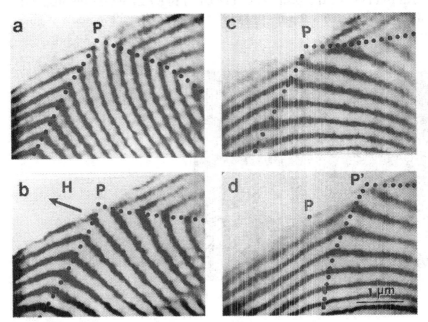

Fig. 7.28a-d. Video scenes showing the dynamics of magnetic domain boundaries in a permalloy thin fim when the magnetic field H changes: (a) H = 0, (b) H = 36 G, and (c) and (d) H = 48 G

domain. When the magnetic field further increased to 48 G, the domain structure suddenly changed from (c) to (d) within one tenth of a second. This can be interpreted that the two domain walls were pinned at the point P, and were depinned to move.

7.10 Static Observation of Fluxons in the Profile Mode

7.10.1 Quantized Flux (Fluxons)

Superconductivity is closely related to magnetic fields in that the superconductive state easily breaks down due to the magnetic field. This is especially true in the case of a **type-I superconductor**. Therefore, large currents cannot be made to flow through this kind of superconductor. However, a **type-II superconductor**, the type to which all practical superconductors such as Nb_3Sn and high-T_c superconducturs belong, has an ingeneous mechanism: an applied magnetic field can pass through the superconductor in the form of extremely thin filaments (Fig. 7.29), thus maintaining superconductivity in the rest of the superconductor. Such a filament of flux, which has a constant magnetic flux of $h/(2e)$ ($=2 \cdot 10^{-15}$ Wb), is called a **fluxon** (or a **vortex**, since the superconductive vortex current flows surrounding the filament). Fluxons play an important role in the fundamentals of superconductivity and in practical applications. For example, the critical current of a superconductor is determined by how tightly the fluxons is attached to some pinning centers when the **Lorentz force** is exerted on them by the current (Fig. 7.29).

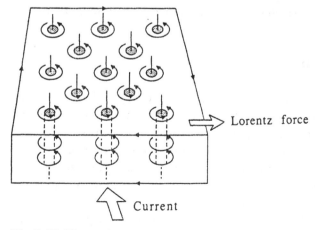

Fig. 7.29. Fluxons in a superconductor. A Lorentz force is exerted on fluxons when a current is applied

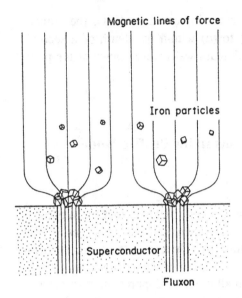

Fig. 7.30. Schematic diagram of the Bitter technique

Several investigators have tried to observe fluxons by using the Bitter method [7.34], Low-temperature scanning electron microscopy [7.35], scanning tunneling microscopy [7.36], scanning Hall probe microscopy [7.37, 38], scanning SQUID microscopy [7.39], and magnetic force microscopy [7.40]. In the Bitter method, for example, magnetic powder is sprinkled on the surface of the superconductor. This powder accumulates at the position of fluxons, forming an image of the fluxons that can be observed using an electron microscope (Fig. 7.30). Nevertheless, the fluxon has evaded the dynamic observation because, in addition to having a very small flux value, it is shaped like an extremely thin filament, perhaps 100 nm in diameter. Electron holography and Lorentz microscopy using a field-emission electron beam have opened up possibilities to directly observe fluxons.

7.10.2 Experimental Method

The experimental method for studying static fluxons [7.41] involves the following procedures. A tungsten wire 30 μm in diameter was used as the substrate for the superconducting specimen. After the wire surface was made clean and smooth by flash heating to 2000 K with an electric current, lead was evaporated onto one side of the wire. The scanning electron micrograph of a sample is shown in Fig. 7.31. The white part in the photograph (b) indicates the Pb film.

In the holography electron microscope this sample was put on a low-temperature specimen stage, and a magnetic field of a few Gauss was ap-

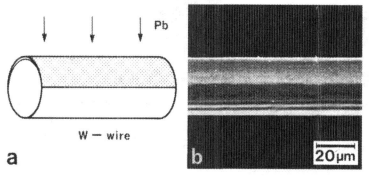

Fig. 7.31a,b. Thin-film evaporated lead onto one side of a tungsten wire: (**a**) Schematic diagram, and (**b**) scanning electron micrograph

plied perpendicular to the film (Fig. 7.32). The specimen was then cooled to 4.5 K. In the weak magnetic field, the Meissner effect excluded the magnetic lines from the superconductor. However, the magnetic lines of force penetrated the superconductor when the magnetic field was strong. An electron beam was incident onto the specimen from above, and the magnetic lines of force were observed through the process of electron holography.

7.10.3 Experimental Results

Figure 7.33 presents the resultant interference micrograph for a superconductive film being 0.2 μm thick. The phase difference in this figure is amplified by a factor of two, so one contour fringe corresponds to one fluxon. A single fluxon was captured on the right-hand side of this photograph. The magnetic line of force was produced from an extremely small area of the lead surface, and then it spreads out into the free space. If we assume an axially symmetric distribution of the fluxon, we can conclude from this photograph that the magnetic field near the surface of the specimen reaches as

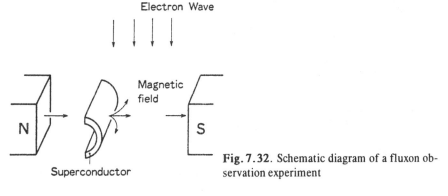

Fig. 7.32. Schematic diagram of a fluxon observation experiment

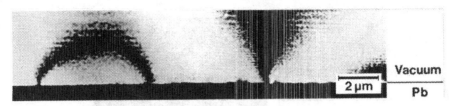

Fig. 7.33. Inteference micrograph of single fluxons (phase amplification: ×2). One fringe corresponds to a flux of h/(2e) (i.e., a single fluxon). A magnetic line of force is produced from the small surface area of a lead film 0.2 μm thick on the right-hand side of the micrograph. An antiparallel pair of fluxons is also seen on the left-hand side

high as 1000 G even though the applied magnetic field is only a few Gauss. This is because a tiny solenoid is formed by the superconducting vortex current (Fig. 7.34), which results in a strong magnetic field.

Although the photograph in Fig. 7.33 provides evidence that the observed magnetic lines of force are the fluxons, further confirmation was obtained by performing the following experiments. When the specimen temperature exceeded the critical temperature, the magnetic lines disappeared. This is because the superconducting state was destroyed, causing the vortex current to stop flowing. This offers proof that the observed magnetic lines were generated by the superconductive current.

Further evidence is that the fluxon corresponded exactly to h/(2e), as it can be concluded from Fig. 7.33. However, the fluxon image is obtained more precisely when it is amplified using phase amplification. Figure 7.35 depicts an interference micrograph in which the image is amplified 16 times. Sixteen interference fringes correspond to the flux of h/e, and eight fringes are observed. This reveals that the quantum flux value is h/(2e) to within 10%. This photograph also gives us a clue to the internal structure of a fluxon. For example, the diameter of this fluxon is 150 nm at the surface.

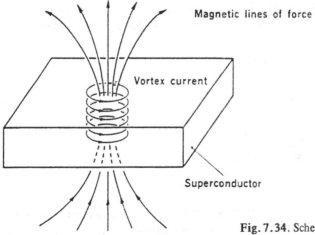

Fig. 7.34. Schematic diagram of a fluxon

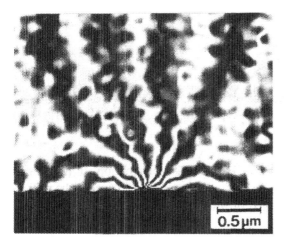

Fig. 7.35. Enlarged interference micrograph of a fluxon (phase amplification: ×16)

If we assume that the fluxon is axially symmetric, we can obtain its magnetic field distribution [7.42].

In addition to isolated fluxons, we also observed a pair of fluxons oriented in opposite directions and connected by magnetic lines of force (Fig. 7.33, left). One possible reason for the pair's creation is as follows: When the specimen is cooled below the critical temperature, lead is brought into the superconductive state. However, during the cooling process the specimen experiences a state in which fluxon pairs appear and disappear repeatedly because of thermal excitations. This phenomenon was predicted by the **Kosterlitz-Thouless theory** [7.43, 44] and is inherent to two-dimensional systems such as thin films or layered structures. It can occur in high-T_c superconductors. During the cooling process, a pair of fluxons are produced, pinned by some imperfections in the superconductor, and eventually frozen. The pair of fluxons presented in Fig. 7.33 were probably created in this way.

What will happen in a thicker superconducting film? Figure 7.36 illustrates the state of the magnetic lines of force when the film is 1 μm thick. We can see that the state is completely changed: the magnetic flux does not penetrate the superconductor individually in fluxon form, but in a bundle. The figure does not indicate any fluxon pairs. The explanation for this phenomena is as follows: Because lead is a *type-I superconductor*, the application of a strong magnetic field partially destroys the superconductive state in some parts of the specimen (**intermediate state**), as depicted in Fig. 7.36. In this photograph, the magnetic lines of force penetrate those parts of the specimen where superconductivity is destroyed, but because the surrounding regions are still superconducting, the total amount of the penetrating magnetic flux is an integral multiple of the flux quantum $h/(2e)$. The

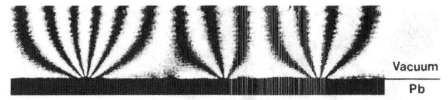

Fig. 7.36. Interference micrograph of fluxons penetrating a lead film 1.0 μm thick (phase amplification: ×2)

thin superconducting film shown in Fig. 7.33 is an exception. In this case, lead behaved like a type-II superconductor and the flux penetrates the superconductor in the form of individual fluxons.

7.11 Dynamic Observation of Fluxons in the Profile Mode

7.11.1 Thermally Excited Fluxons

The magnetic lines of force of a single fluxon can be directly observed by electron holography. Because the flux itself, and not its replica, is visualized with this method, even a dynamic detection of fluxons is feasible. This was actually accomplished by *Matsuda* et al. [7.45], who dynamically observed an electron hologram through a television system and recorded it on videotape (Fig. 7.37). The video signal from the tape was digitized and

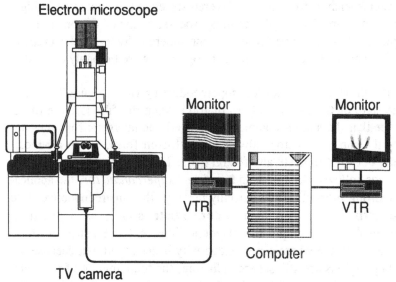

Fig. 7.37. Schematic diagram of dynamic observation of fluxons

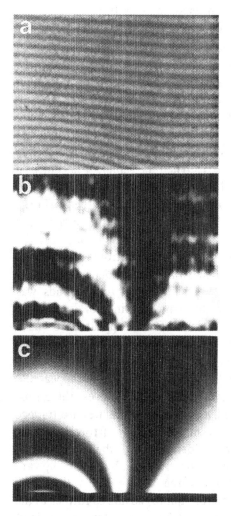

Fig. 7.38a-c. Interference micrographs of flux reconstructed from a hologram recorded on videotape: (**a**) Hologram, (**b**) interference micrograph (phase amplification: × 2), and (**c**) processed interference micrograph (phase amplification: ×2)

stored in a memory device and then transported frame by frame to the computer, which numerically computed the electron phase distribution from the hologram recorded in each frame. This distribution was calculated by using the Fourier-transform method and was displayed as a twofold phase-amplified contour map in which one contour corresponds to the magnetic line of force from a single fluxon, $h/(2e)$.

Examples of a hologram recorded in a videotape frame and of the reconstructed contour map are shown in Fig. 7.38. However, the quality of the contour map (b) is not as good as that of conventional maps reconstructed from holograms that have been recorded on film. This is inevitable for two reasons. First, the exposure time to take an electron hologram is only 1/30 s (much shorter than the few seconds taken for film exposure). Second, the number of carrier fringes in the hologram is as small as 10 to 50, an order of magnitude fewer than in a standard hologram recorded on film.

To eliminate deteriorations such as those due to Fresnel fringes which are produced from biprism filament edges, we made use of the fact that the object here consisted of magnetic fields and that magnetic fields in vacuum cause phase distributions with harmonic functional shapes [7.46]. The resultant contour map is exhibited in Fig. 7.38c. The experimental procedure for generating this map was as follows: a magnetic field of a few Gauss was applied to a superconducting Pb film prepared in the same manner that was used to prepare the film shown in Fig. 7.31. When the sample was cooled to 5 K, the magnetic fluxes were trapped in fluxon bundles by the superconducting lead film. When the applied magnetic field was turned off, the trapped fluxes remained stationary at 5 K. When the sample temperature was raised, the diameters of the fluxes gradually increased and, near the critical temperature, the fluxes began to move. An electron beam was incident on the fluxes and their dynamic behavior was continuously recorded on video-tape for 10 to 20 min. A flux change of a few seconds was selected and reconstructed numerically the magnetic lines of force. The flux changes were spontaneous and various: fluxes appeared to move abruptly from one pinning center to another, and to approach and return between two pinning centers; antiparallel pairs of fluxes were attracted until they finally vanished. When T exceeded T_c, all trapped fluxes disappeared.

Figure 7.39 reveals how the thermally excited flux behaves: three fluxons in the upward direction (indicated by arrows in the figure) are trapped in the center of micrograph (a), and three magnetic lines of force leak into the vacuum. At t = 0.13 s, the fluxons suddenly shift to the left corner of the micrograph (b), and it can be seen in this micrograph that two upward fluxons are connected by magnetic lines of force. At t = 0.70 s, the downward fluxon at the right-hand side of micrograph (b) advances further to the right to disappear in micrograph (c). At t = 1.33 s, an antiparallel pair of fluxons disappears and only one single upward fluxon remains, thereby producing a broad magnetic line. Strictly speaking, since the flux change is completed within 0.03 s (a single frame interval) the behavioral dynamics of any specific flux cannot be followed. This technique is still off-line and does not have a time resolution high enough to resolve the fluxon movement. Development of brighter electron beams and faster image processing techniques will help to achieve real-time dynamic observation.

7.11.2 Current-Driven Fluxons

When the magnetic flux trapped in a superconductor is dynamically observed under thermal excitation, it moves randomly and its movement

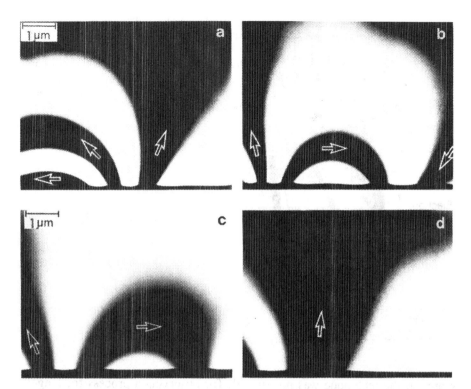

Fig. 7.39a-d. Dynamic flux motion due to thermal activation. When sample temperature increases from 5 K to 7 K, trapped fluxes begin moving, then disappear: (**a**) t = 0 s, (**b**) t = 0.13 s, (**c**) t = 0.70 s, and (**d**) t = 1.33 s

cannot be predicted. But when a current is applied to the superconductor, the Lorentz force builds up proportionally to the current in a direction perpendicular to the flux. Furthermore, the pinning force at each site can be determined by measuring the current on which the flux depends. *Yoshida* et al. [7.47] actually observed current-driven dynamics of trapped fluxons by using real-time interferometry.

a) Experimental Method

The time resolution of the observation technique for studying the dynamics described in the previous section was limited by that of the TV system. Therefore, this technique could not observe a dynamic phenomenon in which the state of the magnetic flux changes more rapidly than the 0.03 s resolution of the recording system, since interference fringes in a two-beam hologram were easily eradicated when multiple exposures corresponding to different flux states overlapped. In addition, a prohibitively large amount of time was required to numerically reconstruct the flux lines from each holo-

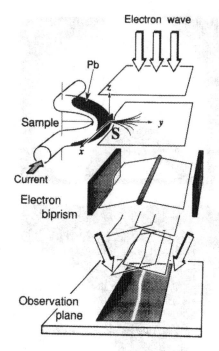

Fig. 7.40. Schematic diagram of the experimental arrangment. Fluxons protruding from the sample are seen as step-shaped kinks in the interference fringe at the location of the flux on the sample surface. The current flowing in the z-direction drives the flux in the $\pm x$ direction, depending on its polarity

gram, so the technique was not suitable for observing a process for longer than approximately one minute.

Therefore, we took advantage of a real-time electron interferometry technique to observe a pinned fluxon as a step-shaped kink in a single interference fringe at a position corresponding to the pinning site (Fig. 7.40). Thus, although in this technique a real image of the fluxons cannot be reconstructed, the sequence of fluxons that are moving but temporarily pinned at a center can be recorded, as an oscillation of the fringe with a node at the position corresponding to the pinning site. The image of magnetic lines of force of fluxons can be reconstructed from a multiple-fringe interference pattern (hologram) obtained simply by increasing the voltage applied to the biprism filament. Figure 7.41 represents an example of the corresponding single-fringe and multiple-fringe interference patterns. The magnitude of the flux can be determined from the multiple-fringe pattern (b). The way in which the fringes are eradicated when the flux state changes frequently is indicated in photograph (c). When the applied current is gradually increased from zero, the whole evolution of the flux dynamics can be traced: from the initial stage of a fluxon cutting into the superconducting strip from its edge, up to the superconductivity breaking down because of rapidly repeated sweeps by the current-driven fluxons.

The sample in this experiment was a tungsten wire 100 μm in diameter and bent to form the intricate shape depicted in Fig. 7.42, enabling an elec-

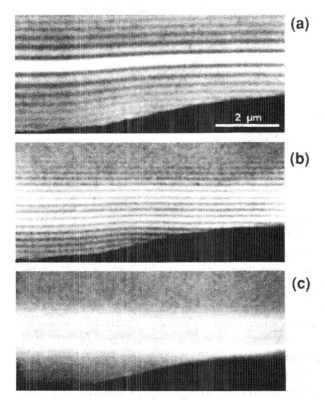

Fig. 7.41a-c. Multiple-fringe interference pattern. The single interference fringe (**a**), which accompanies Fresnel fringes, can be changed to a multiple-fringe pattern (**b**) simply by increasing the voltage applied to the biprism filament to measure the flux. However, the fringes in a multiple-fringe pattern are easily wiped out, as shown in (**c**) when the flux state changes more rapidly than the time resolution of the recording system

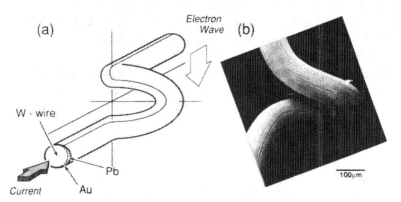

Fig. 7.42a,b. Superconducting film sample: (**a**) Schematic, and (**b**) scanning electron micrograph

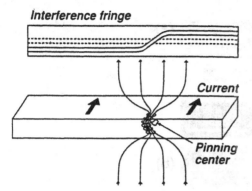

Fig. 7.43. Interference fringe and magnetic flux. The fringe is straight when there is no flux. When the flux is trapped, the fringe has a step-shaped kink at the flux location

tron beam to be orthogonal both to the fluxon direction and to its current-driven motion. A Pb film 0.5 to 1.0 μm thick was evaporated onto one side of the wire. With the sample temperature fixed at 7 K, a collimated electron beam illuminated the sample and then passed through both an electron biprism and an intermediate lens (not shown in Fig. 7.40) to form an image. It consists of a single interference fringe and a shadow of the sample surface (Fig. 7.41a). The image is further enlarged by magnifying electron lenses so that it can be observed via a TV camera and recorded on videotape.

Before showing the experimental results, let us use Fig. 7.43 to consider the relationship between trapped flux and the fringe displacement. The step-shaped kink in the interference fringe implies that flux lines are protruding from the Pb surface. In fact, when there is a flux of magnitude n times that of the fluxon Φ_0 ($= \frac{1}{2}h/e = 2 \cdot 10^{-15}$ Wb), the fringe has a step-shaped kink at a point that corresponds to the flux location. This is because that part of the electron wave passing close to S (Fig. 7.40) is phase shifted by $n\pi$ on both sides of the flux location. This displaces the two portions of the fringe in opposite directions, resulting in the step-shaped kink.

When bulky, lead is a type-I superconductor and does not allow flux to penetrate in the form of fluxons. But when it is thinner than 0.5 μm, it is of type II. Our Pb film was 0.5 to 1.0 μm thick, but the surface regions were probably oxidized and its edges were likely grainy because of the oblique evaporation. Consequently, the film could behave as a type-II superconductor.

Fig. 7.44a-g. Interference fringes when an increasing current is applied to a superconducting lead film: (a) no flux trapped at j = 600 A/mm^2; (b) $-2\Phi_0$ flux pinned in the center at j = 650 A/mm^2; (c) $-\Phi_0$ flux pinned at the right-hand side at j = 650 A/mm^2; (d) $+2\Phi_0$ flux pinned at the left-hand side at j = 650 A/mm^2; (e) rapid flux flow momentarily stopping at the center at j = 700 A/mm^2; (f) rapid flux flow without stopping at the pinning center at j = 720 A/mm^2; (g) no flux trapped due to the breakdown of superconductivity at j = 760 A/mm^2

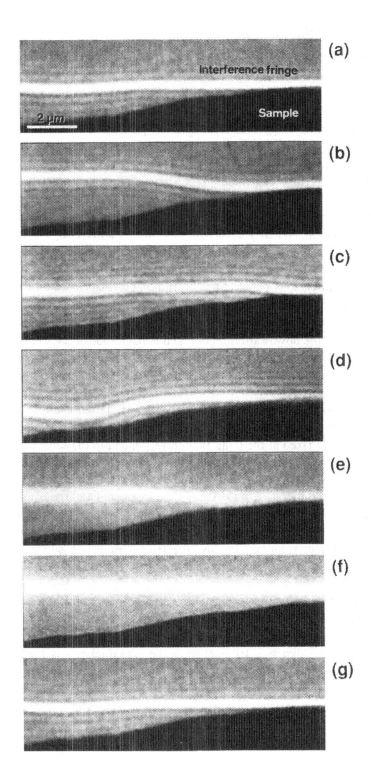

b) Experimental Results

The experimental results presented in Fig. 7.44 have been reproduced from a videotape recording of a process in which the current increased gradually from 0 to 760 A/mm^2. In each photograph, the black part at the bottom is the shadow image of the sample surface, and the white string just above it is the electron-interference fringe. The weak lines parallel to this string, which are seen clearly in Figs. 7.44a-d and g, are Fresnel fringes caused by the biprism wire edges.

The interference fringe remained straight until the current density j reached 650 A/mm^2 (a). Because the small fringe distortion on the left-hand side remained unchanged below and above T_c, it was probably due to the sample charging up. When j exceeded 650 A/mm^2, the fringe bent abruptly to have a step-shaped kink, indicating that the current-induced flux around the superconductor started nucleating to cut into the film from its edges and resulted in fluxons penetrating the film. The fluxons were then driven by the Lorentz force and were trapped at a pinning center in the middle of the field of view (b). The fringe displacement shows that the flux had a polarity pointing in the downward direction and a magnitude of $2\Phi_0$. On a few occasions, fluxes were found to be trapped at two sites on the right-hand and left-hand sides, each bundle having $-\Phi_0$ (c) and $2\Phi_0$ (d), respectively.

When the current reached 700 A/mm^2, the fringe started oscillating very rapidly between straight and bent lines (a and b). The oscillation exhibited in (e) has a node fixed at the point that corresponded to the same pinning center, implying that the fluxons moved more frequently than once per 0.03 s and were momentarily pinned at the center. However, when the current has been increased to 720 A/mm^2 the node disappeared and the whole interference fringe oscillated horizontally many times within 0.03 s (f). This is because a strengthened Lorentz force caused the fluxons to pass by the pinning site without stopping. Finally, the fringe oscillation suddenly stopped at j = 760 A/mm^2 (g) indicating that the superconducting state had broken down.

The pinning force, which has attracted much attention from the viewpoint of both fundamental physics and applications, can be measured for individual sites. For the situation illustrated in Fig. 7.44, the pinning force was found to be $3 \cdot 10^{-7}$ dyn = $3 \cdot 10^{-12}$ N when the film thickness is 1 μm, the magnitude of the flux bundle is $2\Phi_0 = 4 \cdot 10^{-15}$ Wb, and the current j is 720 A/mm^2. This value for a thin Pb film is 2 to 3 times larger than the values extrapolated from the data obtained for Nb by *Park* et al. [7.48] or for Pb-Bi by *Hyun* et al. [7.49].

7.12 Observation of Fluxons in the Transmission Mode

In the profile mode described so far, magnetic lines of force of fluxons leaking from a superconductor surface can be observed by impinging an electron beam parallel to its surface. However, only the magnetic lines of force outside a superconductor can be observed. One cannot visualize fluxon patterns inside a superconductor, interaction of fluxons with structural defects, or two-dimensional fluxon arrays.

A transmission mode has to be adopted to provide for these possibilities. Observation of fluxons with transmission electron microscopes was attempted [7.50-55] but could not be carried out because of the lack of a coherent and penetrating electron beam until recently, when the 350-kV electron-holographic microscope was developed.

7.12.1 Experimental Methods

In the transmission mode, the superconducting thin film has to be tilted, as shown in Fig. 7.45. The reason for this is obvious. Since only components of magnetic fields perpendicular to the electron-beam direction can interact with the electron beam, the phase shift caused by a single fluxon becomes too small to be detected when the electron beam impinges perpendicularly to the film plane and consequently parallel to a fluxon [7.54]. The diffraction contrast produced when the electron incidence perpendicular to a fluxon was calculated by *Capiluppi* et al. [7.55].

In addition, the film sample has to be uniform in thickness. Even a small variation in film thickness may produce an electron phase change that would hide a fluxon phase shift as small as approximately $\pi/2$. One may

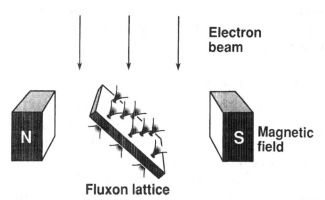

Fig. 7.45. Experimental arrangement for observing the fluxon lattice. A superconducting thin film is tilted so that electrons may interact with fluxons' magnetic fields

think that the phase shift caused by a single fluxon should be π irrespective of the tilting angle, but this is not the case. Because of the Meissner effect for the tilted superconducting film, the phase shift is smaller than π, depending on the tilt angle, the film thickness and the fluxon diameter, as was shown by numerical simulation [7.56, 57].

Individual fluxons can be seen by both interference microscopy and Lorentz microscopy: In an interference micrograph, the projected magnetic lines of force are visualized as contour fringes drawn on the in-focus electron micrograph. Sample structures are observed with high spatial resolution and, at the same time, magnetic lines of force are quantitatively measured. With Lorentz microscopy, a fluxon can be seen as a pair of bright and dark contrast features. This method, though not high in resolution, is convenient for real-time observation of the fluxon dynamics.

7.12.2 Experimental Results

Individual fluxons have recently been observed in the transmission mode using both a Lorentz microscope [7.58] and an interference microscope [7.59]. As illustrated in Fig. 7.46a,b, the specimen arrangement is the same for these two kinds of microscopy but the observation methods are different.

A single-crystalline Nb thin film was prepared by chemically etching a roll-film. The film, set on a low-temperature stage, was tilted at 45° to the incident electrons impinging vertically so that the electrons could be affected by the fluxons' magnetic fields which penetrates the film perpendicular to its surface. An external magnetic field of up to 150 Gauss was applied to the film in the horizontal direction.

Fluxon information is contained in the phase distribution or the wavefront distortion of the transmitted electron beam. This cannot be read out from a conventional electron micrograph where only the intensity distribution is recorded. However, the phase distribution can be measured directly in a holographic interference micrograph.

a) Behavior of Fluxons in Nb Thin Films

An example of an interference micrograph for fluxons is displayed in Fig. 7.47. Projected magnetic lines of force can directly be observed as contour fringes in the micrograph being phase-amplified 16 times. The regions where magnetic lines become locally dense correspond to fluxons.

Simple image defocusing can make fluxons visible, and thus the dynamic behavior of fluxons can be observed. Phase modulation caused by fluxons reveals itself in a Lorentz micrograph. Here, a fluxon can be seen

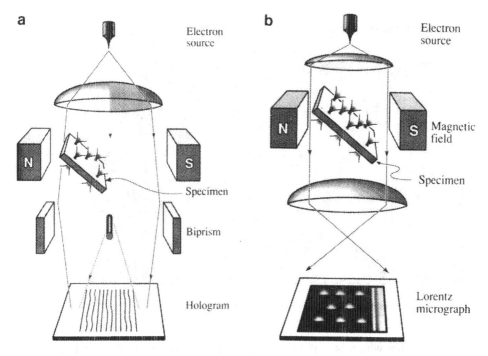

Fig. 7.46a,b. Observation methods for observing fluxon lattice: (**a**) Electron-holographic interference microscopy, and (**b**) Lorentz microscopy. Since fluxons are a phase object to an illuminating electron beam, they become observable either by measuring the phase shift (**a**) or by defocusing the image (**b**)

Fig. 7.47. Interference micrograph of a fluxon lattice in a superconducting Nb film (phase amplification: ×16). Encircled regions in the photograph correspond to fluxons

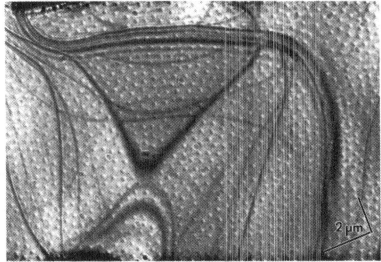

Fig. 7.48. Lorentz micrograph of a fluxon lattice in a superconducting Nb film. Each spot with black and white contrast pair corresponds to a fluxon

as a tiny spot which is half bright and half dark. An example of a Lorentz micrograph is illustrated in Fig. 7.48. Each black-and-white spot indicates a fluxon. The line dividing the two contrast regions indicates the direction of each fluxon. Since the black part appears on the same side for all the spots, the polarities of all the fluxons are the same. When the applied magnetic field was reversed, this contrast reverses, too, as expected.

These two photographs show static images of fluxons. The dynamic behaviour of the individual fluxons was observed for the first time using Lorentz microscopy and recorded on a videotape. An example is displayed below.

The sample was first cooled down to 4.5 K and the applied magnetic field B was gradually increased. No fluxons were observed until B reached 32 G, when a few fluxons suddenly came into the field of view. The number of fluxons increased with an increase in B. Their dynamic behavior is quite noteworthy. After the magnetic field and the sample temperature changed somewhat, the fluxons continued to move for a few minutes as if they were searching for an equilibrium state. The fluxons oscillated around their own pinning centers and occasionally jumped from one pinning center to another one as long as they were not packed too densely ($B \leq 100 G$). When B suddenly changed from 100 G to zero, 90% of the fluxons instantly disappeared. The remaining fluxons gradually moved towards a hole in the film, jumping among the pinning centers (Fig. 7.49). This phenomenon seems to be due to the gradient of the film thickness, and also due to the repulsive force among trapped vortices in the film.

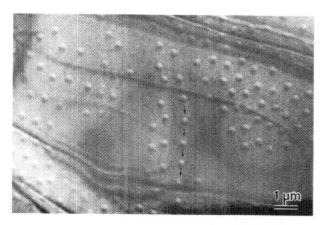

Fig. 7.49. Lorentz micrograph of fluxons in a Nb thin film when B is changed abruptly from 100 to 0 G. Fluxons begin to hop from time to time downward the edge of the film just like jumping over stepping stones

Fluxons behave in a very interesting manner especially when there exist strong pinning centers. Since such behavior is significant not only from a scientific viewpoint but also from practical ones, new results that have been obtained by Lorentz microscopy are introduced here. The reason is as follows: A superconductor is a material to which an electric current can be applied without any dissipation. However, this is possible only when fluxons remain resistant against the Lorentz force exerted on them by the current. The developments of superconducting materials with large critical currents have, in essence, been made by trial and error, and the mechanism of the flux pinning has not been fully elucidated. Lorentz microscopy allows us to directly observe the microscopic scenary of the flux pinning.

b) Estimation of Pinning Forces of Defects

The **pinning force** of individual defects was initially estimated by dynamically observing fluxons when the applied magnetic field changed. Several lines of defects were produced by the irradiation of a focussed ion beam perpendicularly to the edge of a Nb film so that fluxons may move along the defect lines. The irradiation dose changed from line to line, and the dependence of the force on the ion dose was investigated.

When a magnetic field of 100 G was applied, fluxons were produced so densely that we could not observe which fluxons were pinned at the defects. When the magnetic field decreased, unpinned fluxons at first began to leave the film. In the meantime, weakly pinned fluxons also began to move. They hopped from one defect to another one along a line of defects just like jumping over stepping stones. Figure 7.50 depicts a Lorentz micrograph of this state. It can be seen from this micrograph that all the unpinned fluxons

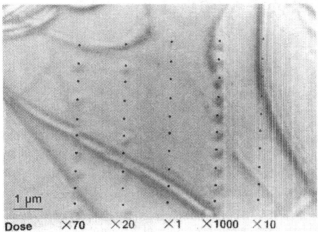

Fig. 7.50. Trapped fluxons at lines of defects even when the applied magnetic field was switched off. The ion doses when the defects were produced, are shown in units of 10^{-10} C

leave the film. Fluxons pinned at defects produced by ion irradiation of 1000 times the unit dose did not move at all during the process of the experiment, where the unit dose was 10^{-10} C. No fluxons remained to be trapped at defects due to 1 and 10 times the unit dose. In the case of defects due to 20 and 70 times the unit doses, some of the pinned fluxons left the film depending on the ion doses stated in the figure. Such observations confirmed that the elementary pinning forces of individual defects can be estimated by observing the fluxon dynamics when a force is exerted on the fluxons; in the present case the pinning force increased with the higher-dose defects.

The critical current of a superconductor cannot be increased by making stronger the pinning force of the defects alone, since unpinned fluxons begin to move easily. Although no fluxons may easily move if every fluxon is pinned one by one by each defect, such a situation is not practical since the number of fluxons soon changes when the magnetic field or the current changes. Therefore, the behavior of fluxons was investigated when pinned fluxons and unpinned fluxons coexisted.

c) Intermittent Rivers of Fluxons

When a weak magnetic field ($<7G$) is applied to a Nb thin fim which has point defects with medium pinning forces (ion dose: $\times 70$), unpinned fluxons being far away from the defect trapping of a fluxon move independently of the existence of the defects. Only fluxons located at, or passing near, defects notice an effect from the pinned fluxons such that the defects trapping fluxons generate some resitance to the flow of unpinned fluxons:

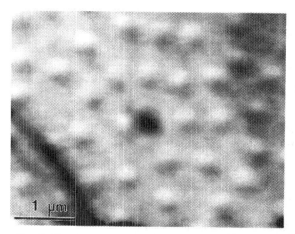

Fig. 7.51. Video frame of flowing fluxons that detour around the black defects

most unpinned fluxons had to go around the black defect which traps a fluxon (Fig. 7.51). Since unpinned fluxons did not move smoothly but hop from one weak pinning site to another one, there is a sudden chance where the fluxons collide with the defect at high speed. In this case, sometimes an additional fluxon enters the defect. However, two trapped fluxons are unstable since the defect radius is comparable to that of a fluxon. Consequently, in a few seconds one of them leaves the defect. On rare occasions, fluxons even bounce off the defect.

For dense fluxons, very interesting phenomena were found to occur [7.60]. Fluxons can be compared to bar magnets and repel each other due to their magnetic fields. When they are squeezed by an external magnetic field, they tend to form a closely packed lattice. Where strong pinning centers exist, fluxons cannot form a lattice, but instead they form domains of lattices that produce domain boundaries (Fig. 7.52).

The question is what happens to fluxons when a force is exerted on such a configuration of fluxons and increases. The result is that fluxons flow intermittently in rivers, like an avalanche, along the domain boundaries. Each defect strongly pins not only a single fluxon but also a bundle of fluxons. When the force reaches a critical value, the weakest regions of the fluxon arrangements along the domain boundaries collapse and the fluxons in that region flow in rivers.

A video frame of such a river of fluxons is depicted in Fig. 7.53. It can be seen that the fluxon image is blurred inside the river since fluxons move during the exposure time (1/30s) of one video frame. This flow stops at 0.5s forming a new configuartion of fluxons. When it comes to the state where fluxons cannot resist, fluxons flow again along new domain boundaries. Such a process is repeated, and as a result, intermittent rivers of fluxons flow here and there. This is the first direct experimental observation

Fig. 7.52. Fluxon configuration in a Nb thin film with defects (T = 7.5 K, B = 75 G). Black spots indicate point defects produced by the irradiation of a focused Ga-ion beam, and white spots indicate fluxons. Fluxons cannot form a single lattice since fluxons are strongly trapped at defects. When you see this micrograph at a grazing angle, you will note domain boundaries

that fluxons begin to flow in the form of a plastic flow [7.61]. Simulations were made by *Reichhardt* et al. [7.62, 63].

d) Matching Effect

The behavior of fluxons is very different, however, when the defects are arranged densely and regularly. It is known that the critical current of such

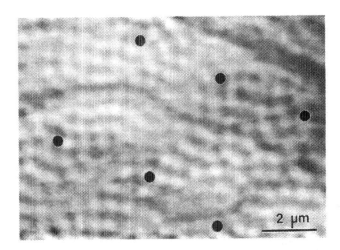

Fig. 7.53. Video frame indicating a river of fluxons. When a force is exerted on fluxons which form domains of lattices due to the existence of defects, the fluxons begin to flow in a river along the domain boundaries. The fluxon images are blurred in a river due to the movement of fluxons in a video frame

a superconductor has peaks at specific values of the applied magnetic field. The microscopic mechanism of this **matching effect** has been investigated by Lorentz microscopy using a Nb thin film with a square array of point defects [7.64]. A configuration of defects is illustrated in Fig. 7.54.

Lorentz micrographs exhibiting the configuration of fluxons in specific magnetic fields are depicted in Fig. 7.55. At the "matching" magnetic field H_1, all the defects are occupied by fluxons without any vacancies, thus forming a square lattice (Fig. 7.55c). When fluxons form a regular and rigid lattice, even if a fluxon is depinned from one pinning site due to thermal excitation, it cannot find any vacant site to move to. As a result, a bigger force is required if these fluxons are to be moved

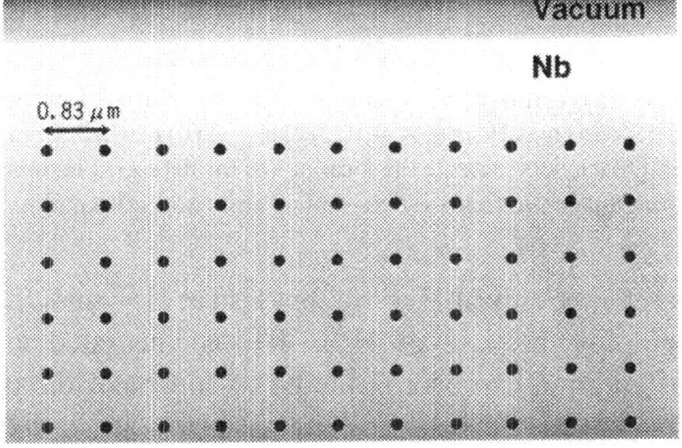

Fig. 7.54. Configuration of point defects produced by irradiation of a focused ion beam

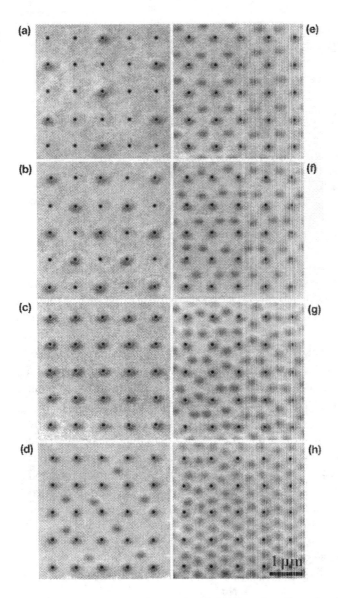

Fig. 7.55a-h. Lorentz micrographs of the fluxon configuration in a square arry of point defects at matching magnetic feldsH$_n$: (**a**) n = 1/4, (**b**) n = 1/2, (**c**) n = 1, (**d**) n = 3/2, (**e**) n = 2, (**g**) n = 3, and (**h**) n = 4. Small black dots indicate the locations of the defects. Fluxons form regular lattices at the matching magnetic field H$_1$, as well as at its multiples and its fractions

Regular lattices are formed not only at H = H$_1$, but also at H = mH$_1$/n (m, n: integers). An example for the case of H = 4H$_1$ is illustrated in Fig. 7.55h. This fluxon configuration can be considered to consist of three kinds of arrangements: First, all the defects forming a square lattice are occupied by fluxons. Then, two fluxons aligned in the vertical direction are inserted at every interstitial site. Finally, an additional fluxon is inserted in the middle of two adjacent defects in the vertical direction. Figure 7.55a displays the case of $H = (1/4)H_1$. Fluxons occupy every fourth site in the horizontal direction, thus forming a centered (4×2) rectangular lattice. The reason for the pinning force as a whole to become stronger at specific magnetic field strengths is that lattices that are formed by fluxons, are rigid and regular. When "excess" or "deficient" fluxons are produced at magnetic fields different from the specific values, the fluxons could be induced to hop by a weaker force, just like "electrons" and "holes" in a semiconductor. When a stronger force is applied to fluxons forming a regular lattice such as that shown in Fig.7.55e in the vertical direction, the flow of fluxons is very different: it looks like a simultaneous movement of the interstitial fluxons along a line.

An apparently strange event was also noted: Although a fluxon may strongly be pinned at a defect and never depinned, the fluxon was *effectively* depinned [7.65]. Such an event can actually happen due to the production of **antifluxons**, the magnetic fields of which are oriented in the opposite direction to those of fluxons. However, the production of antifluxons has often been ignored since the magnetic fields of a fluxon and an antifluxon cancel each other so far as macroscopic measurements are concerned.

Antifluxons are found to be produced even in a very simple experiment. When the magnetic field applied to a niobium thin film with a hole increases as high as 10 G, fluxons and antifluxons are both produced in the regions A and C near the hole edge, respectively, as illustrated in Fig.7.56a. The numbers of fluxons and antifluxons are the same. Neither fluxons nor antifluxons are found in the regions B and D. The reason why antifluxons were produced whose magnetic field is in the opposite direction of the applied field cannot be explained.

Our possible interpretation is as follows: Since the thin region near the hole edge is completely surrounded by a thick film region, fluxons cannot easily penetrate the film from the outside. If fluxons happen to be produced by some asymmetry of the film near the hole, the same number of antifluxons have to be produced.

When the magnetic field increases further, the numbers of fluxons and antifluxons both increase, and the regions B and D narrow to finally vanish at 50 G (Fig.7.56b). At higher magnetic fields, the number of fluxons increases while the number of antifluxons decreases (Fig.7.56c). This is because fluxons begin to penetrate the sample from the outside, and antifluxons start to disappear by pair annihilation.

The explanation described above has been confirmed by the following experiment: When the film was cut, as depicted in Fig. 7.56d, no antifluxons were produced. Fluxons have already begun to appear above 1 G.

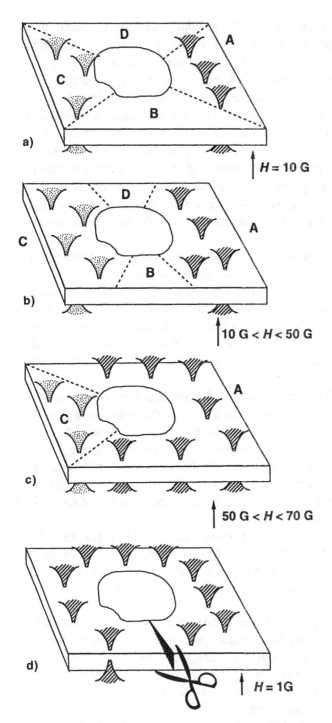

Fig. 7.56a-d. Production of fluxons and antifluxons by the application of a magnetic field to a Nb thin film with a hole: (**a**) H = 10 G, (**b**) 10 G < H < 50 G, (**c**) 50 G < H < 70 G, and (**d**) H = 1 G

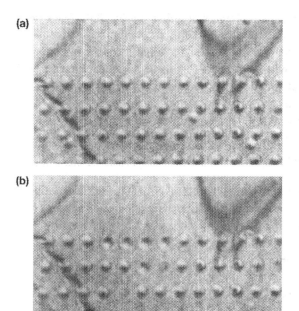

Fig. 7.57a,b. Trapped fluxons in Nb thin film with an array of point defects when the magnetic field decreases from the matching field H_1: (**a**) $H = 0.9H_1$ and (**b**) $H = 0.6H_1$. Even when the magnetic field decreases to $0.9H_1$, fluxons outside the defect region disappear. When $H = 0.6H_1$, antifluxons are produced from the film edge and collide head-on with trapped vortices to disappear

As this example suggests, antifluxons often appear and exhibit an apparently strange behavior which can never be predicted from macroscopic magnetization measurements. There are also cases where antifluxons produce an effect on the flux pinning. Such an example is represented in Fig. 7.57. A magnetic field is applied to a Nb thin film which has a square array of artificial point defects. When the matching magnetic field H_1 of 30 Gauss is reached, fluxons are produced whose density is the same as that of these defects. Therefore, all the defects are occupied by fluxons in the defect region. When the magnetic field decreases, fluxons outside a defect region begin to leave the film. Already at $H = 0.9H_1$, all the fluxons outside the defect region have left the film, whereas all the fluxons inside it remained to be trapped (Fig. 7.57a).

When the magnetic field further decreases, trapped fluxons do not start to depin. Instead, antifluxons are produced at the film edge, approach the defect region, and collide head-on with trapped fluxons to disappear. The reason for the penetration of antifluxons is as follows: The magnetic line of a trapped fluxon produced from the top surface of the film goes a long way beyond the film edge and returns to the original fluxon from the back surface. Since the magnetic field is applied in the opposite direction at the film edge, antifluxons are produced.

The annihilation of a trapped fluxon with an incoming antifluxon is equivalent to the depinning of a trapped fluxon. The depinning can take place, no matter how strong the pinning force of a defect may be. When an incoming antifluxon collides with a trapped fluxon to disappear, a vacant site is produced at the defect location. Subsequent antifluxons arriving at vacant sites are trapped there as they arrive.

The Lorentz micrograph in Fig. 7.57b shows the case of $H = 0.6H_1$. The black side of the fluxon image is on the lower side, while that of the antifluxon image is on the upper side. Some of the defects in the front row are occupied by antifluxons.

Such dynamic interaction between fluxons and antifluxons could not be detected with macroscopic measurements, since their magnetic fields cancel each other. The observation by Lorentz microscopy revealed that antifluxons are often produced and annihilated with fluxons which produce an effect on flux pinning that cannot be ignored.

e) High-T_c Superconductors

A high-T_c superconductor has been investigated by means of Lorentz microscopy [7.66]. The main reason why high-T_c superconductors have not been developed to practical use as expected is that the critical current vanishes at a temperature even well below the critical temperature T_c. This is attributed to the behaviour of fluxons, but details of the mechanism have not yet been clearly understood. Most scientists believe that fluxons melt like molecules in a liquid, thus making flux-pinning difficult at pinning sites [7.67]. In this case the temperature for practical use would not be T_c but rather the melting temperature T_m. Evidence for flux melting was reported using a Bitter figure of fluxons in $Bi_2 Sr_{1.8} CaCu_2 O_x$ (BSCCO), which was

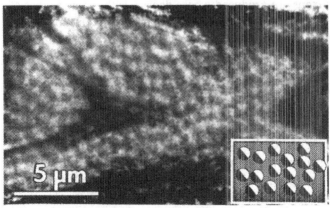

Fig. 7.58. Lorentz micrograph of fluxons in a $Bi_2 Sr_{1.8} CaCuO_x$ thin film at $B = 20$ G and $T = 4.5$ K

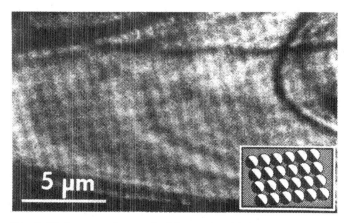

Fig. 7.59. Lorentz micrograph of fluxons in a $Bi_2Sr_{1.8}CaCuO_x$ thin film at B = 20 G and T = 56 K

obtained by an electron-microscopy observation of the decorated pattern of the superconductor surface with small ferromagnetic particles. This hypothesis was concluded from the blurred fluxon image even at 15 K [7.68]. Others, however, disagreed with this interpretation and attributed the phenomenon to weak pinning effects.

Fluxons in a BSCCO film were observed to investigate until at what temperature fluxons do not begin to move. The experiment was made with the sample subjected to a fixed magnetic field B and a temperature increasing from 4.5 K to above T_c. The Lorentz micrograph at T = 4.5 K and B = 20 Gauss is shown in Fig. 7.58. Fluxons are randomly distributed. When the temperature was raised stepwise by a few K, the fluxons moved. After a few minutes, however, the fluxons arrived at an equilibrium state and turned stationary. They did not melt even at 20 K. The fluxon configuration began to change at 40 K: Fluxons formed a regular lattice above this temperature due to the weakened pinning effect (Fig. 7.59). The fluxon lattice persisted at higher temperatures, although the image contrast gradually decreased and then disappeared above 77 K.

Recently, the fluxon dynamics in a BSCCO film has been investigated experimentally [7.69]. When the magnetic field was weak and consequently fluxons were sparsely distributed, the motion of fluxons was found to be different above and below 25 K, the value of which changed a little bit depending on the sample under investigation. Below 25 K, all the fluxons moved slowly as if they were moving in a viscous fluid. While above 25 K, fluxons hopped suddenly from one site to another. The slow migration can be interpreted to be due to the collective pinning of a fluxon due to a large number of oxygen defects. When the temperature increases, the pinning force of tiny oxygen defects decreases rapidly and larger defects become dominant.

These two kinds of fluxon movements can be seen in the Internet of the Nature Home Page (http://www.nature.com/cgi-bin/SupplData.cgi/author/Tonomura.+A.)

Other fluxon movements can also be observed:
Vortex river (http://www.aaas.org/science/matsuda.htm)
Matching effect (http:www.sciencemag.org/feature/data/harada.shl).

8. High-Resolution Microscopy

The original objective of holography was to improve the resolution of an electron microscope. Since an aberration-free system cannot be achieved by the combination of concave and convex lenses in the electron case because of the lack of a concave electron lens, the resolution is determined by the aberrations of the objective lens and not by a fundamental limit of the electron wavelength. In fact, it was theoretically verified by *Scherzer* [8.1] that an axially-symmetric magnetic field can function only as a convex lens for an electron beam.

This chapter first explains how seriously the spherical aberration affects a high-resolution image and then introduces experimental attempts to holographically correct the spherical aberrations.

8.1 Phase Contrast Due to Aberration and Defocusing

In high-resolution electron microscopy, a specimen can often be regarded as a weak phase object. That is, the object wave ϕ_0 just behind the object is expressed as

$$\phi_0 = e^{i\psi} \ . \tag{8.1}$$

When $|\psi| \ll 2\pi$,

$$\phi_0 \approx 1 + i\psi \ . \tag{8.2}$$

In this case, an in-focus image formed with an ideal lens has no contrast because there are no intensity variations, i.e., $|\phi_0|^2 = 1$. Therefore an image is usually defocused so that in the widest domain of spatial frequency, both spherical aberration and defocusing can be used to produce a relative phase difference of $\pi/2$ between the transmitted wave and the scattered wave. Under this condition, the phase distribution is converted to the intensity distribution [8.2,3]. Since the first and second terms in (8.2) correspond to the transmitted and scattered waves, respectively, the resultant image amplitude

ϕ_0 is influenced by the effects of the lens aberration and defocusing. It becomes

$$\phi_0' = 1 + i\psi e^{-\pi i/2} = 1 + \psi, \tag{8.3}$$

thus producing the intensity variations

$$|\phi_0'|^2 \approx 1 + 2\psi. \tag{8.4}$$

However, this phase contrast cannot be produced in the wide region of spatial frequency of the image: if an appropriated defocusing condition is not selected, only a special spatial frequency in the image is emphasized.

Scherzer [8.4] investigated the conditions under which the phase contrast is produced in the widest domain of the spatial frequency to get a high-resolution image. His findings are roughly explained as follows: The transmittance of an object can be considered to consist of sinusoidal gratings of different periodicities and different directions. An electron beam incident onto each grating is diffracted under the angle α being inversely proportional to the grating spacing d ($\alpha = \lambda/d$). The diffracted beam consequently passes through a different part of the objective lens. As a result, the diffracted beam experiences a phase shift depending on the grating spacing due to the spherical aberration of the lens that results in a displacement of the image of a sinusoidal grating (Fig. 8.1). As the grating spacing becomes narrower, the displacement of the grating image becomes larger.

In general, the phase modification $W(\alpha)$ (Sect. 5.2.5) of imaging electrons during their passage through the objective lens is given by

$$W(\alpha) = \frac{2\pi}{\lambda}\left[\frac{C_s}{4}\alpha^4 - \frac{\Delta f}{2}\alpha^2\right], \tag{8.5}$$

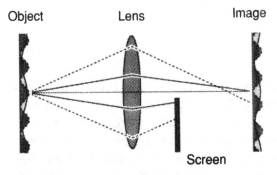

Fig. 8.1. Effect of the spherical aberration on sinusoidal objects

where C_s is the spherical-aberration constant of the objective lens and Δf is the defocusing distance. If the wave-front aberration $W(\alpha)$ takes on the extreme value $-\pi/2$ at $\alpha = \alpha_0$, that is, if $W'(\alpha_0) = 0$, then the following relation holds:

$$\alpha_0 = (\lambda/C_s)^{1/4}, \quad \text{and} \quad \Delta f = (C_s\lambda)^{1/2}. \tag{8.6,7}$$

Under this condition (Fig. 8.2c), $W(\alpha)$ is negative until α reaches α_1:

$$\alpha_1 = (4\lambda/C_s)^{1/4}. \tag{8.8}$$

When $\alpha > \alpha_1$, the aberration $W(\alpha)$ has a positive value and increases rapidly with α.

The condition expressed by (8.4) holds only for $\alpha = \alpha_0$. For a general value of α, the value of $|\phi'_0|^2$ is given by

$$|\phi'_0|^2 = 1 - 2\psi \sin W(\alpha). \tag{8.9}$$

Here, according to *Hanszen* [8.5], the phase-contrast transfer function $C_p(\alpha)$ can be defined as a multiplication factor when the phase variation is transformed into an intensity variation. At $\alpha = \alpha_0$, $C_p(\alpha)$ is equal to 2. When $\alpha \neq \alpha_0$, the function $C_p(\alpha)$ is given by

$$C_p(\alpha) = -2 \sin W(\alpha) \tag{8.10}$$

and is plotted in Fig. 8.2. Figure 8.2c shows that the contrast of a phase grating disappears at $\alpha = \alpha_1$ (or at the grating spacing of λ/α_1) and the contrast reverses its sign frequently with further increasing α.

The point resolution d is defined as the half-width of the point-image diffraction-limited by the objective aperture, which selects only the frequency domain where the contrast transfer function has a positive value. The resolution is given, in the case of the coherent illumination, by

$$d = 0.8\lambda/\alpha_1 = 0.6 C_s^{1/4} \lambda^{3/4}. \tag{8.11}$$

The resolution becomes high when the spherical aberration constant decreases, but the phase-contrast transfer function $C_p(\alpha)$ vanishes when C_s becomes zero at $\Delta f = 0$.

All the discussions up to now are based on the assumption that the wave-front aberration $W(\alpha)$ takes on the extremum value of $-\pi/2$ at $\alpha = (\lambda/C_s)^{1/4}$ for $\Delta f = (C_s\lambda)^{1/2}$. However, it would be possible to obtain a higher resolution if this condition is removed. The aberration $W(\alpha)$ is also plotted against α for various values of the defocusing distance Δf in

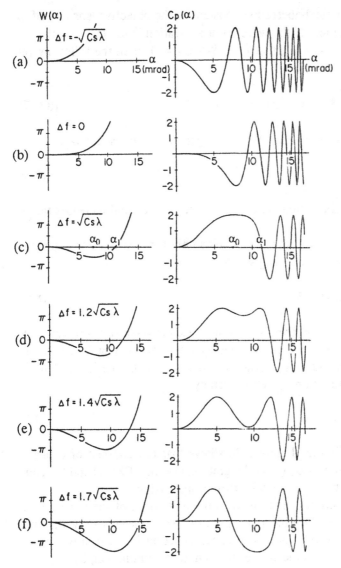

Fig. 8.2a-f. Wave-front aberration and phase contrast transfer function ($C_s = 1$mm, $\lambda = 3.7$pm): (a) $\Delta f = -(C_s\lambda)^{1/2}$; (b) $\Delta f = 0$; (c) $\Delta f = (C_s\lambda)^{1/2}$; (d) $\Delta f = (3/(2\pi)^{1/2})(C_s\lambda)^{1/2}$; (e) $\Delta f = 1.4(C_s\lambda)^{1/2}$; (f) $\Delta f = 1.7(C_s\lambda)^{1/2}$

Fig. 8.2. It can be seen in Fig. 8.2a that when $\Delta f < 0$ (**overfocusing**), $W(\alpha)$ increases monotonically with increasing α and has no extremum. The resolution defined in (8.11) becomes lower with increasing $|\Delta f|$ because α_1, at which $C_p(\alpha)$ crosses the horizontal axis, decreases. However, when $\Delta f > 0$ (**underfocusing**), $W(\alpha)$ always has an extremum. For example, as already described, $C_p(\alpha)$ has an extremum of $-\pi/2$ for $\Delta f = (C_s\lambda)^{1/2}$ (Fig. 8.2c). When Δf increases a little bit from $(C_s\lambda)^{1/2}$, the value of α_1 increases but

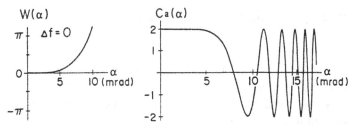

Fig. 8.3. Wave-front aberration and amplitude transfer function for $\Delta f = 0$

the contrast $|C_p(\alpha)|$ at the extremum weakens instead. *Scherzer* [8.4] adopted $\Delta f = (3/(2\pi))^{1/2}(C_s \lambda)^{1/2}$ as the optimum condition (Fig. 8.2d). Under this condition, $W(\alpha)$ is $(1/2 - 3/(2\pi)^{1/2})\pi \approx -0.7\pi$ at $\alpha = \alpha_0$, and consequently $C_p(\alpha_0) = 1.6$. When Δf further increases, $C_p(\alpha)$ begins to oscillate for smaller α. It should be noted here that when $\Delta f = 1.7(C_s \lambda)^{1/2}$ (Fig. 8.2f), $C_p(\alpha)$ has the first peak at $\alpha = 4.6$ mrad and then the broad contrast band around $\alpha = 10.5$ mrad appears. This condition is for the observation of the lattice image, which will be discussed in Sect. 8.2.2.

For a weak amplitude object, the amplitude contrast transfer function can be obtained by similar considerations (Fig. 8.3):

$$C_a(\alpha) = 2\cos W(\alpha) . \qquad (8.12)$$

The contrast vanishes at an angle where the scattered wave experiences a phase shift of $-\pi/2$, and the amplitude object is consequently regarded as a phase object. A contrast inversion occurs at larger angles. In this case, the contrast does not vanish even at $C_s = 0$ because the object is an amplitude object.

The electron phase can directly be derived from an electron hologram without recourse to the effect of defocusing and aberration. The contrast transfer function of the phase image thus obtained is different from that given by (8.10). For example, when the spherical aberration is completely compensated for in the reconstruction stage of holography, this contrast-transfer function remains a constant value over all the region of α. More precisely, the contrast-transfer function of the phase image turns into the same form as that for an amplitude object as given by (8.12). In this way, the holographically reconstructed phase image, whose spherical aberration is ideally compensated for, displays a contrast proportional to the phase distribution even under the in-focus condition.

8.2 Optical Correction of Spherical Aberration

An experiment was first carried out by *Tonomura* et al. to optically compensate for the spherical aberration in holography [8.6]. This section introduces their experiments mainly to help explain the physical meaning of spherical aberration and its compensation.

8.2.1 In-Focus Electron Micrograph of a Crystalline Particle

The effect of spherical aberration can often be observed when we observe high-resolution images of crystalline objects. An example of an in-focus electron micrograph of a fine gold particle is depicted in Fig. 8.4. This particle is not a single crystal but a **multiply-twinned particle** [8.7, 8], and consequently the condition of an incident beam for exciting Bragg reflections is different for each single-crystal domain. The dark contrast in the upper part of the particle image implies then that an incident electron beam is reflected by (111)-lattice planes, thereby producing a dark contrast in the transmitted image. The two beams reflected by the lattice planes form images at different positions because of the spherical aberration of the objective lens under the in-focus condition. In fact, two white regions can be seen on both sides of the transmitted particle image in the photograph. They are images formed with such Bragg-reflected beams and, because of the spherical aberration, they are as much as 10 nm apart from the transmitted image. Inside these two regions, lattice fringes with a 240 pm spacing can be observed. These fringes are formed by the interference between the incident beam

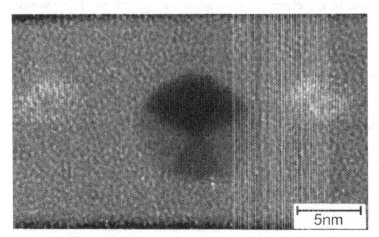

Fig. 8.4. In-focus electron micrograph of a fine gold particle

and the reflected beams, but they are *virtual* in the sense that no crystals exist at that location.

Lattice images are not usually observed under the in-focus condition because such virtual images are seen. They must therefore be underfocused so that lattice fringes may be formed at the proper position. However, in this case, the image is blurred because of the defocusing required. In addition, lattice fringes with different spacings cannot be formed at the same location because the defocusing conditions differ for different Bragg-reflected angles.

8.2.2 Off-Axis Hologram of a Crystalline Particle

An example of a hologram is presented in Fig. 8.5. The image is greatly defocused so that Bragg-reflected images may be formed inside the transmitted image. The defocusing distance required for this condition is much larger than that for *Scherzer*'s condition: the Bragg-reflected beam and the transmitted beam overlap when the image is defocused by $C_s \alpha^2 = C_s (\lambda/d)^2$ = 600 nm, whereas the defocusing distance is 100 nm for *Scherzer*'s condition given by $\Delta f = (3/(2\pi)^{1/2})(C_s \lambda)^{1/2}$. Consequently, the outline of the particle image is blurred in this image.

Lattice fringes whose spacings are shorter than the resolution of the measurement can often be observed in electron microscopy. In the above experiment, for example, lattice fringes of 240 pm and even 120 pm can be observed, whereas the point resolution calculated from (8.11) is 400 pm (λ

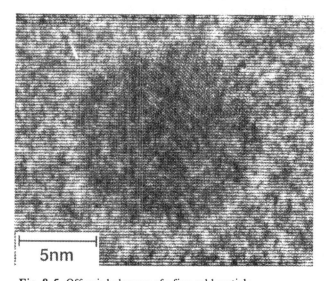

Fig. 8.5. Off-axis hologram of a fine gold particle

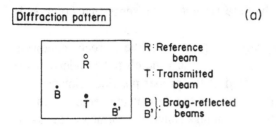

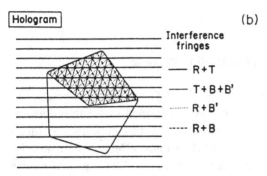

Fig. 8.6a,b. Interpretation of the interference fringes in the hologram shown in Fig. 8.5: (**a**) Electron diffraction pattern, and (**b**) schematic diagram of interference fringes

= 4.5pm, C_s = 1.7mm). This situation can be understood from the contrast-transfer function shown in Fig. 8.2f. The contrast is high around the spatial frequency α = 10 mrad corresponding to a 240-pm spacing. In addition, $C_p(\alpha)$ has a broad peak at that spatial frequency. However, at lower spatial frequencies the contrast is reversed.

In Fig. 8.5, various kinds of interference fringes are observed in the region where the transmitted and the reflected images overlap. Lattice fringes in the vertical direction are fringes formed by the interference between the transmitted beam T and the Bragg-reflected beam B and B' (Fig. 8.6b). The interference fringes between the two reflected beams B and B' also appear where their intensities are high in the central region. Biprism fringes formed by the interference between the transmitted beam T and the reference beam R are observed in the horizontal direction all over the field of view inside and outside the particle image. Additionally, there are two kinds of oblique fringes that are formed by interference between the reference beam R and the two reflected beams B and B'. These oblique fringes are important because they are evidence that the interference fringes between the reference beam and all the scattered beams are recorded in a hologram.

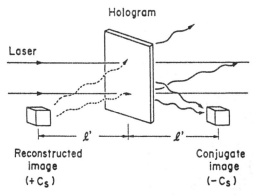

Fig. 8.7. Reconstruction of twin images. The reconstructed image is affected by the spherical aberration C_s, whereas the conjugate image by $-C_s$

8.2.3 Image Reconstruction

An optically reconstructed image can be formed in one of the two diffracted beams by illuminating the hologram with a laser beam, as depicted in Fig. 8.7. This image is affected by the spherical aberration C_s of the electron lens. This is because only the scattered beam and not the reference beam is phase shifted by the aberrations of the objective lens in the electron microscope, whereas the conjugate image formed in the other diffracted beam undergoes the spherical aberration with the opposite sign $-C_s$. This can be understood if we recall that the phase distribution of the conjugate image is reversed in sign compared with that of the reconstructed image.

An optically reconstructed image is depicted in Fig. 8.8. This image is in focus, but images under arbitrary focusing conditions can be obtained simply by defocusing this image on an optical bench. The positions of the Bragg-reflected beams move with the defocusing distances, and it was confirmed that Bragg-reflected beams could be reconstructed optically.

8.2.4 Spherical Aberration Correction

An optical system compensating for the spherical aberration of a reconstructed image is presented in Fig. 5.16. In this experiment, the conjugate image, which is subjected to spherical aberration of $-C_s$, is corrected by the $+C_s$ spherical aberration of the convex lens. The resultant corrected image is illustrated in Fig. 8.9. Even under the in-focus condition, Bragg-reflected images are formed inside the transmitted image of the particle. Half-spacing fringes are observed in the inner region of the particle, where the intensity of the transmitted beam vanishes and only the interference pat-

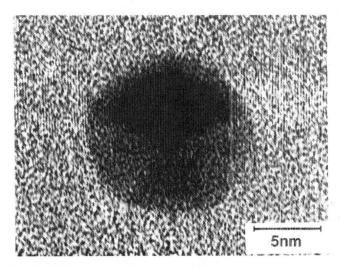

Fig. 8.8. Reconstructed image of a fine gold particle under the in-focus condition. Because of the spherical aberration, the lattice images formed with Bragg-reflected beams are reconstructed 10 nm far from the transmitted image

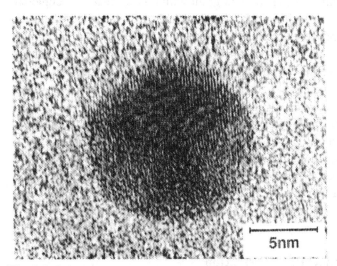

Fig. 8.9. Spherically corrected image of a fine gold particle

tern between the two reflected beams B and B' (Fig. 8.6) occurs. When the additional image reconstructed in the other diffracted beam is observed under the same condition, that image is subjected to twice the spherical aberration $+2C_s$, since the aberration C_s of the reconstructed image further increased by $+C_s$ because of the correction lens. This was confirmed by observing the distance between the Bragg-reflected image and the transmitted image is twice as large as that in Fig. 8.8 under the in-focus condition. In this way, the spherical aberration was demonstrated to be compensated for within a precision of about 1/10 of C_s.

8.3 Numerical Correction of the Spherical Aberration

Lichte et al. [8.9] numerically reconstructed structure images of Nb_2O_5 crystals from their high-resolution holograms and, as shown in Fig.8.10, they also eliminated the effect of spherical aberration. A structure image of a crystalline object differs from a lattice image in that it displays a projected atomic structure observed from the direction of an atomic array and is formed with many diffracted beams, whereas only a few lattice fringes are formed in the lattice image. The structure image is therefore an interference pattern formed by using many diffracted beams. Generally, the more beams used, the more faithful the image becomes.

As many as 150 reconstructed diffraction beams were used when the hologram was formed with a 350-kV field-emission electron microscope [8.10]. The optical diffraction pattern of its hologram is presented in Fig. 8.11. Two sidebands appear on both sides of the center band. The center band represents the optical diffraction pattern of the electron micrograph, and the two sidebands describe the electron diffraction patterns. The center band extends to a radius of two thirds of the distance L between the center and one sideband. An image is reconstructed from one sideband region, whose radius is L/3. The resolution of a holographically reconstructed im-

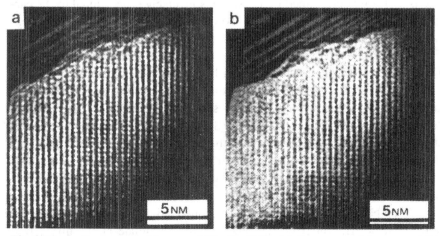

Fig. 8.10a,b. Reconstructed images of Nb_2O_5 [8.9]: (**a**) Reconstructed image, and (**b**) spherically corrected image

Fig. 8.11. Optical diffraction pattern of a hologram of H-Nb$_2$O$_5$ [8.10]

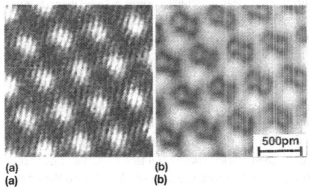

(a) (b)
(a) (b)

Fig. 8.12a,b. Silicon thin film [8.11]: (a) Hologram and (b) aberration-corrected amplitude image

age is therefore determined by L, which is inversely proportional to the carrier-fringe spacing of a hologram. To obtain a 100 pm resolution, for example, the carrier fringe spacing should be less than 30 pm (Sects. 4.2, 5.1.1). The shortest spacing reported so far for off-axis holograms is 9 pm [8.10], so a resolution much finer than 100 pm can be expected.

Of course, this is not the only condition for getting a high-resolution image. As discussed in Sect. 8.1, the spherical aberration of the electron lens introduces the phase shift $W(\alpha)$, (8.5), to the amplitude of the electron diffraction pattern. Therefore, only the diffracted beam of $\alpha < \alpha_1$ (Fig. 8.2c) should be used to reconstruct a faithful image of an object. Sinusoidal images formed with diffracted beams of $\alpha > \alpha_1$ are not only shifted, but sometimes they also have no contrast or reversed contrast.

A higher-resolution image can be obtained by first subtracting the wave-front aberration from the phase of the amplitude at the diffraction plane, and then reconstructing the image.

In fact, *Orchowski* et al. improved the image resolution by this process and observed the dumbbell structure of a Si thin film (Fig.8.12) [8.11]. More recently, *Lang* et al. recorded the dumbbell atom images of CdTe and ZnTe revealing different intensities due to the difference in atomic numbers [8.12]. A still different method to improve the image resolution from a focal series of images (in-line holograms) was developed by *Coene* et al. [8.13].

9. Conclusions

Electron holography was invented by D. Gabor in 1948, but new applications of electron holography have only recently been opened up by making the best use of coherent beams from both a laser and a field-emission electron beam. Holography faithfully transforms electron wave fronts into optical wave fronts, enabling versatile optical techniques to become applicable to electron optics. The original objective of holography was to exceed the resolution limit of electron microscopes by compensating, in the optical reconstruction stage, for the spherical aberration of the electron lens. Attempts have been made to actually achieve this objective, thus increasing the levels of technology to where we are on the brink of improving the resolution and making new discoveries.

When interferometric techniques of optical holography are used in the optical reconstruction stage of electron holography, the phase distribution of the electron wave function in an image plane can be determined quantitatively. In fact, the precision in phase measurement has reached the order of 1/100 of the electron wavelength, and this high-precision electron interferometry has made it possible for us to quantitatively observe thickness distribution in atomic dimensions, the magnetic lines of force of a ferromagnetic thin film, and even individual fluxons penetrating a superconductor. Applications are not restricted only to such high-precision measurements, but also extend to experimental investigations in fundamental physics, such as the Aharonov-Bohm effect.

Although the experimental progress has a long history, applications have been devised only recently. It can be said that electron holography is just at the starting line, with many potential applications ahead. Further developments both in basic electron holography techniques and in their applications are expected along with the development of a more *coherent* and more *penetrating* electron beam, that is, a higher-voltage and brighter electron beam. I hope many researchers enter into this fascinating field of physics.

References

Chapter 1

1.1 D. Gabor: Microscopy by reconstructed wavefronts. Proc. Roy. Soc. (London) A **197**, 454 (1949)
1.2 D. Gabor: Microscopy by reconstructed wavefronts II. Proc. Phys. Soc. (London) B **64**, 449 (1951)
1.3 E.N. Leith, J. Upatnieks: Reconstructed wavefronts and communication theory. J. Opt. Soc. Am. **52**, 1123 (1962)
1.4 A. Tonomura, T. Matsuda, J. Endo, H. Todokoro, T. Komoda: Development of a field emission electron microscope. J. Electron. Microsc. **28**, 1 (1979)
1.5 E. Zeitler: Electron holography, in *Proc. Electron Microscopy Soc. of America*, ed. by G.W. Bailey (Claitor, Baton Rouge 1979) pp.376-379
1.6 G.F. Missiroli, G. Pozzi, U. Valdrè: Electron interferometry and interference electron microscopy. J. Phys. E **14**, 649 (1981)
1.7 K.-J. Hanszen: Holography in electron microscopy, in *Advances in Electronics and Electron Physics*, Vol.59, ed. by L. Marton (Academic, New York 1982) pp.1-77
1.8 H. Lichte: Electron holography. Proc. 10th Int'l Congress on Electron Microscopy, Hamburg 1982 (Deutsche Gesellschaft für Elektronenmikroskopie, Frankfurt 1982) Vol.1, pp.411-418
1.9 A. Tonomura: Applications of electron holography. Rev. Mod. Phys. **59**, 637 (1987)
1.10 A. Tonomura: Present and future of electron holography. J. Electron Microsc. **38**, S43 (1989)
1.11 A. Tonomura: Electron holography, a new view of the microscopic. Phys. Today **22**, 22 (1990)
1.12 A. Howie: Electrons give a broader view. Nature **345**, 386 (31 May 1990)
1.13 H. Lichte: Electron image plane off-axis holography of atomic structures, in *Advances in Optical and Electron Microscopy*, ed. by T. Mulvey, C.J.R. Sheppard (Academic, London 1991) Vol.12, pp.25-91
1.14 P.W. Hawkes, E. Kasper: *Principles of Electron Optics*, Vol.3, *Wave Optics* (Academic, London 1994)

Chapter 2

2.1 R. Meier: Magnification and third-order aberrations in holography. J. Opt. Soc. Am. **55**, 987 (1965)

2.2 E.N. Leith, J. Upatnieks: Reconstructed wavefronts and communication theory. J. Opt. Soc. Am. **52**, 1123 (1962)

2.3 J.B. DeVelis, G.B. Parrent, B.J. Thompson: Image reconstruction with Fraunhofer holograms. J. Opt. Soc. Am. **56**, 423 (1966)

Chapter 3

3.1 L. Reimer: *Transmission Electron Microscopy*, 4th edn., Springer Ser. Opt. Sci., Vol. 36 (Springer, Berlin, Heidelberg 1997)

3.2 T. Hibi: Pointed filaments I. Its production and its application. J. Electron Microsc. **4**, 10 (1956)

3.3 A.N. Broers: Some experimental and estimated characteristics of the lanthanum hexaboride rod cathode electron gun. J. Sci. Instrum. **2**, 273 (1969)

3.4 A.V. Crewe, D.N. Eggenberger, D.N. Wall, L.M. Welter: Electron gun using a field emission source. Rev. Sci. Instrum. **39**, 576 (1968)

3.5 P.W. Hawkes (ed.): *Magnetic Electron Lenses*, Topics Curr. Phys., Vol. 18 (Springer, Berlin, Heidelberg 1982)

3.6 G. Möllenstedt, H. Düker: Beobachtungen und Messungen an Biprisma-Interferenzen mit Elektronenwellen. Z. Phys. **145**, 377 (1956)

3.7 T. Hibi, K. Yada: Electron interference microscopy, in *Principles and Techniques of Electron Microscopy*, Vol. 6, ed. by M.A. Hayat (Van Nostrand, New York 1976) pp. 312-343

3.8 G. Matteucci: On the use of a Wollaston wire in a Möllenstedt-Düker electron biprism. Microsc. et Spectrosc. Electronique **2**, 69 (1978)

3.9 A. Tonomura, J. Endo, T. Matsuda, T. Kawasaki, H. Ezawa: Demonstration of single-electron buildup of an interference pattern. Am. J. Phys. **57**, 117 (1989)

3.10 P.W. Hawkes: Coherence in electron optics, in *Adv. Opt. and Electron Microsc.* **7**, 101-184 (Academic, London 1978)

3.11 H. Boersch: Experimentelle Bestimmung der Energieverteilung in thermisch ausgelösten Elektronenstrahlen. Z. Phys. **139**, 115 (1954)

3.12 G. Möllenstedt, G. Wohland: Direct interferometric measurement of the coherence length of an electron wave packet using a Wien filter. Proc. Eur. Cong. Electron Microscopy (The Hague 1980) ed. by P. Brederoo, J. Van Landuyt (7th Europ. Congr. on Electron Microscopy Foundation, Leiden 1980) Vol. 1, pp. 28-29

3.13 H. Schmid: Coherence length measurement by producing extremely high phase shifts. Proc. Europ. Congr. Electron Microscopy, Budapest, 1984, ed. by A. Csanady, P. Rohlich, D. Szabo (Program Committee, Budapest 1984) Vol. 1, pp. 285-286

3.14 M. Nicklaus, F. Hasselbach: Wien filter: A wave-packet-shifting device for restoring longitudinal coherence in charged-matter-wave interferometers. Phys. Rev. A **48**, 152 (1993)

Chapter 4

4.1 D. Gabor: Microscopy by reconstructed wavefronts. Proc. Roy. Soc. London A **197**, 454 (1949)

4.2 M.E. Haine, J. Dyson: A modification to Gabor's proposed diffraction microscope. Nature **166**, 315 (19 August 1950)

4.3 H.-W. Fink, W. Stocker, H. Schmid: Holography with low energy electrons. Phys. Rev. Lett. **65**, 1204 (1990)

4.4 H.-W. Fink, H. Schmid: Atomic resolution in lensless low-energy electron holography. Phys. Rev. Lett. **67**, 1543 (1991)

4.5 H. Boersch: Fresnelsche Beugungserscheinungen im Übermikroskop. Naturwiss. **28**, 711 (1940)

4.6 G.I. Rogers: Experiments in diffraction microscopy. Proc. Roy. Soc. Edinburgh A **63**, 193 (1950/51)

4.7 A.V. Baez: A study in diffraction microscopy with special reference to X-rays. J. Opt. Soc. Am. **42**, 756 (1952)

4.8 P. Kirkpatrick, H.M. El-Sum: Image formation by reconstructed wavefronts I. Physical principles and methods of refinements. J. Opt. Soc. Am. **46**, 825 (1956)

4.9 M.E. Haine, T. Mulvey: The formation of the diffraction image with electrons in the Gabor diffraction microscope. J. Opt. Soc. Am. **42**, 763 (1952)

4.10 T. Hibi: Pointed filaments I. Its production and its application. J. Electron Microsc. **4**, 10 (1956)

4.11 E.N. Leith, J. Upatnieks: Reconstructed wavefronts and communication theory. J. Opt. Soc. Am. **52**, 1123 (1962)

4.12 J.B. DeVelis, G.B. Parrent, B.J. Thompson: Image reconstruction with Fraunhofer holograms. J. Opt. Soc. Am. **56**, 423 (1966)

4.13 A. Tonomura, A. Fukuhara, H. Watanabe, T. Komoda: Optical reconstruction of image from Fraunhofer electron-hologram. Jpn. J. Appl. Phys. **7**, 295 (1968)

4.14 K.-J. Hanszen: Holographische Rekonstruktions-Verfahren in der Elektronenmikroskopie und ihre kontrastübertragunstheoretische Deutung. Teil A: In-line Fresnel-Holographie. Optik **32**, 74 (1970)

4.15 J. Munch: Experimental electron holography. Optik **43**, 79 (1975)

4.16 M. Bonnet, M. Troyon, P. Gallion: Possible applications of Fraunhofer holography in high resolution electron microscopy. Proc. Int'l Congress on Electron Microscopy, Toronto, 1978, ed. by J.M. Sturgess (Microscopical Society of Canada, Toronto 1978) Vol. 1, pp. 222-223

4.17 G. Möllenstedt, H. Wahl: Elektronenholographie und Rekonstruktion mit Laserlicht. Naturwissenschaften **55**, 340 (1968)

4.18 A. Tonomura: Electron beam holography. J. Electron Microsc. **18**, 77 (1969)

4.19 I. Weingärtner, W. Mirandé, E. Menzel: Enhancement of resolution in electron microscopy by image holography. Optik **30**, 318 (1969)

4.20 H. Tomita, T. Matsuda, T. Komoda: Electron microholography by two-beam method. Jpn. J. Appl. Phys. **9**, 719 (1970)

4.21 H. Tomita, T. Matsuda, T. Komoda: Off-axis electron microholography. Jpn. J. Appl. Phys. **11**, 143 (1972)

4.22 G. Saxon: Division of wavefront side-band Fresnel holography with electrons. Optik **35**, 195 (1972)

4.23 G. Saxon: The compensation of magnetic lens wavefront aberrations in side-band holography with electrons. Optik **35**, 359 (1972)

4.24 A. Tonomura, T. Matsuda, J. Endo, H. Todokoro, T. Komoda: Development of a field emission electron microscope. J. Electron Microsc. **28**, 1 (1979)

4.25 A. Tonomura, T. Matsuda, J. Endo: High resolution electron holography with field emission electron microscope. Jpn. J. Appl. Phys. **18**, 9 (1979)

4.26 H. Lichte: Electron biprism interference fringes of 0.08 nm spacing for high resolution electron holography. Optik **70**, 176 (1985)

4.27 H. Lichte: Electron holography approaching atomic resolution. Ultramicroscopy **20**, 293 (1986)

4.28 E. Völkl, H. Lichte: Electron holograms for subangstrom point resolution. Ultramicroscopy **32**, 177 (1990)

4.29 T. Kawasaki, T. Matsuda, J. Endo, A. Tonomura: Observation of a 0.055 nm spacing lattice image in gold using a field emission electron microscope. Jpn. J. Appl. Phys. **29**, L508 (1990)

4.30 T. Kawasaki, Q. Ru, T. Matsuda, Y. Bando, A. Tonomura: High resolution holography observation of $H-Nb_2O_5$. Jpn. J. Appl. Phys. **30**, L1830 (1991)

4.31 T. Tanji, K. Urata, K. Ishizuka, Q. Ru, A. Tonomura: Observation of atomic surface potential by electron holography. Ultramicroscopy **49**, 259 (1993)

4.32 G. Matteucci, G.F. Missiroli, G. Pozzi: A new off-axis Fresnel holographic method in transmission electron microscopy. Ultramicroscopy **8**, 403 (1982)

4.33 Q. Ru, N. Osakabe, J. Endo, A. Tonomura: Electron holography available in a non-biprism transmission electron microscope. Ultramicrosc. **53**, 1 (1994)

4.34 Q. Ru: Incoherent electron holography. J. Appl. Phys. **77**, 1421 (1995)

4.35 R. Lauer: Fourier-Holographie mit Elektronen. Optik **67**, 159 (1984)

Chapter 5

5.1 L. Reimer: *Transmission Electron Microscopy*, 4th edn., Springer Ser. Opt. Sci., Vol. 36 (Springer, Berlin, Heidelberg 1997)

5.2 G. Möllenstedt, H. Düker: Beobachtung und Messungen an Biprisma-Interferenzen mit Elektronen Wellen. Z. Phys. **145**, 377 (1956)

5.3 A. Tonomura: Optimum design of accelerating electrodes for a field emission electron gun. Jpn. J. Appl. Phys. **12**, 1065 (1973)

5.4 K. Matsumoto, M. Takahashi: Phase-difference amplification by nonlinear holograms. J. Opt. Soc. Am. **60**, 30 (1970)

5.5 J. Endo, T. Matsuda, A. Tonomura: Interference electron microscopy by means of holography. Jpn. J. Appl. Phys. **18**, 2291 (1979)

5.6 J. Endo, T. Kawasaki, T. Osakabe, A. Tonomura: Sensitivity improvement in electron-holographic interferometry. Proc. 13th Int'l Commission for Optics (Sapporo 1984), ed. by H. Ohzu (Organizing Committee, Sapporo 1984) pp. 480-481

5.7 A. Tonomura, T. Matsuda, T. Kawasaki, J. Endo, N. Osakabe: Sensitivity-enhanced electron-holographic interferometry and thickness-measurement applications at atomic scale, Phys. Rev. Lett. **54**, 60 (1985)

5.8 H.I. Bjelkhagen: *Silver-Halide Recording Materials and Their Processing*, Springer Ser. Opt. Sci., Vol. 66 (Springer, Berlin, Heidelberg 1993)

5.9 M. Takeda, H. Ina, S. Kobayashi: Fourier-transform method of fringe pattern analysis for computer-based holography and interferometry. J. Opt. Soc. Am. **72**, 156 (1982)
5.10 M. Takeda, Q. Ru: Computer-based highly sensitive electron-wave interferometry. Appl. Opt. **24**, 3068 (1985)
5.11 J.H. Bruning: Fringe scanning interferometers, in *Optical Shop Testing*, ed. by D. Malacara (Wiley, New York 1978) pp.409-437
5.12 T. Yatagai, K. Ohmura, S. Iwasaki, S. Hasegawa, J. Endo. A. Tonomura: Quantitative phase analysis in electron holographic interferometry. Appl. Opt. **26**, 377 (1987)
5.13 S. Hasegawa, T. Kawasaki, J. Endo, A. Tonomura: Sensitivity-enhanced electron holography and its application to magnetic recording investigations. J. Appl. Phys. **65**, 2000 (1989)
5.14 Q. Ru, J. Endo, T. Tanji, A. Tonomura: Phase-shifting electron holography by beam tilting. Appl. Phys. Lett. **59**, 2372 (1991)
5.15 G. Lai, T. Hirayama, K. Ishizuka, T. Tanji, Tonomura: Three-dimensional reconstruction of electric-potential distribution in electron-holographic interferometry. Appl. Opt. **33**, 829 (1994)
5.16 G. Lai, T. Hirayama, A. Fukuhara, K. Ishizuka, T. Tanji, A. Tonomura: Three-dimensional reconstruction of magnetic vector fields using electron-holographic interferometry. J. Appl. Phys. **75**, 4593 (1994)
5.17 J. Chen, T. Hirayama, G. Lai, T. Tanji, K. Ishizuka, A. Tonomura: Real-time electron-holographic interference microscopy with a liquid-crystal spatial light modulator. Opt. Lett. **18**, 1887 (1993)
5.18 J. Chen, G. Lai, K. Ishizuka, A. Tonomura: Method of compensating for aberrations in electron holography by using a liquid-crystal spatial-light modulator. Appl. Opt. **33**, 1187 (1994)
5.19 O. Scherzer: The theoretical resolution limit of the electron microscopy. J. Appl. Phys. **20**, 20 (1949)
5.20 G. Saxon: The compensation of magnetic lens wavefront aberrations in side-band holography with electrons. Optik **35**, 359 (1972)
5.21 A. Tonomura, T. Matsuda, J. Endo: Spherical-aberration correction of electron lens by holography. Jpn. J. Appl. Phys. **18**, 1373 (1979)
5.22 D. Gabor: Microscopy by reconstructed wavefronts. Proc. Roy. Soc. London A **197**, 454 (1949)
5.23 I. Weingärtner, W. Mirandé, E. Menzel: Enhancement of resolution in electron microscopy. Optik **30**, 318 (1969)
5.24 M. Haider, S. Uhlemann, E. Schwan, H. Rose, B. Kabius, K. Urban: Electron microscopy image enhanced. Nature **392**, 768 (28 April 1998)
5.25 A. Tonomura, T. Matsuda: A new method for micro-area electron diffraction by electron holography. Jpn. Appl. Phys. **19**, L97 (1980)

Chapter 6

6.1 Y. Aharonov, D. Bohm: Significance of electromagnetic potentials in quantum theory. Phys. Rev. **115**, 485 (1959)
6.2 M. Peshkin, A. Tonomura: *The Aharonov-Bohm Effect*, Lect. Notes Phys., Vol. 340 (Springer, Berlin, Heidelberg 1989)

6.3 T.T. Wu, C.N. Yang: Concept of nonintegrable phase factors and global formulation of gauge fields. Phys. Rev. D **12**, 3845 (1975)
6.4 C.N. Yang: Vector potential, gauge field and connection on a fiber bundle, in *Quantum Coherence and Decoherence*, Proc. 5th Int'l Symp. Foundations of Quantum Mechanics in Light of New Technologies, Tokyo 1995 (Elsevier, Amsterdam 1996) pp. 307-314
6.5 J.C. Maxwell: *A Treatise on Electricity and Magnetism* (Dover, New York 1954) Vol. 2, Article 540, p. 187
6.6 J.C. Maxwell: *A Treatise on Electricity and Magnetism* (Dover, New York 1954) Vol. 2, Article 590, p. 232
6.7 C.N. Yang: *Hermann Weyl's Contribution to Physics. Hermann Weyl 1885-1985*, ed. by K. Chandrasekharan (Springer, Berlin, Heidelberg 1986) pp. 7-21
6.8 C.N. Yang, R.L. Mills: Conservation of isotopic spin and isotopic gauge invariance. Phys. Rev. **96**, 191 (1956)
6.9 R. Utiyama: Invariant theoretical interpretation of interaction. Phys. Rev. **101**, 1597 (1956)
6.10 S. Weinberg: A model of leptons. Phys. Rev. Lett. **19**, 1264 (1967)
6.11 A. Salam: Weak and electromagnetic interactions. Proc. 8th Int'l Symp. on Elementary Particle Theory, ed. by N. Svartholm (Almqvist & Wilksells, Stockholm 1968) pp. 367-377
6.12 H.J. Bernstein, A.V. Philips: Fiber bundles and quantum theory. Sci. Am. **245**, 95 (July 1981)
6.13 R.G. Chambers: Shift of an electron interference pattern by enclosed magnetic flux. Phys. Rev. Lett. **5**, 3 (1960)
6.14 H.A. Fowler, L. Marton, J.A. Simpson, J.A. Suddeth: Electron interferometer studies of iron whiskers. J. Appl. Phys. **32**, 1153 (1961)
6.15 H. Boersch, H. Hamisch, K. Grohmann: Experimenteller Nachweis der Phasenverschiebung von Elektronenwellen durch das magnetische Vektorpotential. Z. Phys. **169**, 263 (1962)
6.16 G. Möllenstedt, W. Bayh: Kontinuierliche Phasenschiebung von Elektronenwellen im kraftfeldfreien Raum durch das magnetische Vektorpotential eines Solenoids. Phys. Bl. **18**, 299 (1962)
6.17 P. Bocchieri, A. Loinger: Nonexistence of the Aharonov-Bohm effect. Nuovo Cimento A **47**, 475 (1978)
6.18 P. Bocchieri, A. Loinger, G. Siragusa: Nonexistence of the Aharonov-Bohm effect. II Discussion of the Experiments. Nuovo Cimento A **51**, 1 (1979)
6.19 D. Home, S. Sengupta: A critical reexamination of the Aharonov-Bohm effect. Am. J. Phys. **51**, 942 (1983)
6.20 S.M. Roy: Condition for nonexistence of Aharonov-Bohm effect. Phys. Rev. Lett. **44**, 111 (1980)
6.21 W. Ehrenberg, R.W. Siday: The refractive index in electron optics and the principles of dynamics. Proc. Phys. Soc. (London) B **62**, 8 (1949)
6.22 D. Bohm, B.J. Hiley: On the Aharonov-Bohm effect. Nuovo Cimento A **52**, 295 (1979)
6.23 T. Takabayasi: Hydrodynamical formalism of quantum mechanics and Aharonov-Bohm effect. Prog. Theor. Phys. **69**, 1323 (1983)
6.24 P. Bocchieri, A. Loinger, G. Siragusa: The role of the electromagnetic potentials in quantum mechanics. The Marton experiment. Nuovo Cimento A **56**, 55 (1980)

6.25 C.G. Kuper: Electromagnetic potential in quantum mechanics: a proposed test of the Aharonov-Bohm effect. Phys. Lett. A **79**, 413 (1980)
6.26 V.L. Lyuboshitz, Ya.A. Smorodinskii: The Aharonov-Bohm effect in a toroidal solenoid. Sov. Phys. - JETP **48**, 19 (1978)
6.27 A. Tonomura, T. Matsuda, R. Suzuki, A. Fukuhara, N. Osakabe, H. Umezaki, J. Endo, K. Shinagawa, Y. Sugita, H. Fujiwara: Observation of the Aharonov-Bohm effect by electron holography. Phys. Rev. Lett. **48**, 1443 (1982)
6.28 A. Tonomura, T. Matsuda, J. Endo, T. Arii, K. Mihama: Direct observation of fine structure of magnetic domain walls by electron holography. Phys. Rev. Lett. **44**, 1430 (1980)
6.29 A. Fukuhara, K. Shinagawa, A. Tonomura, H. Fujiwara: Electron holography and magnetic specimens. Phys. Rev. B **27**, 1839 (1983)
6.30 P. Bocchieri, A. Loinger, G. Siragusa: Remarks on "Observation of Aharonov-Bohm effect by electron holography". Lett. Nuovo Cimento **35**, 370 (1982)
6.31 H. Miyazawa: Quantum mechanics in a multiply-connected region. Proc. 10th Hawaii Conf. High Energy Physics, Hawaii (University of Hawaii 1985) pp. 441-458
6.32 E. Schrödinger: Die Mehrdeutigkeit der Wellenfunktion. Ann. Phys. **32**, 49 (1938)
6.33 W. Pauli: Über ein Kriterium für Ein- oder Zweiwertigkeit der Eigenfunktionen in der Wellenmechanik. Helv. Phys. Acta **12**, 147 (1939)
6.34 A. Tonomura, N. Osakabe, T. Matsuda, T. Kawasaki, J. Endo, S. Yano, H. Yamada: Evidence for the Aharonov-Bohm effect with magnetic field completely shielded from electron wave. Phys. Rev. Lett. **56**, 792 (1986)
6.35 A. Tonomura, S. Yano, N. Osakabe, T. Matsuda, H. Yamada, T. Kawasaki, J. Endo: Proof of the Aharonov-Bohm effect with completely shielded magnetic field. Proc. 2nd Int'l Symp. Foundations of Quantum Mechanics, Tokyo 1986, ed. by M. Namiki, Y. Ohnuki, Y. Murayama, S. Nomura (Phys. Soc. Jpn., Tokyo 1987) pp. 97-105
6.36 N. Osakabe, T. Matsuda, T. Kawasaki, J. Endo, A. Tonomura, S. Yano, H. Yamada: Experimental confirmation of the Aharonov-Bohm effect using a toroidal magnetic field confined by a superconductor. Phys. Rev. A **34**, 815 (1986)

Chapter 7

7.1 I. Sunagawa: Step height of spirals on natural hematite crystals. Am. Mineral. **46**, 1216 (1961)
7.2 J. Endo, T. Kawasaki, T. Matsuda, N. Osakabe, A. Tonomura: Sensitivity improvement in electron holographic interferometry. Proc. 13th Int'l Commission for Optics. Sapporo, 1984, ed. by H. Ohzu (Organizing Committee of ICO-13, Sapporo 1984) pp. 480-481
7.3 A. Tonomura, T. Matsuda, T. Kawasaki, J. Endo, N. Osakabe: Sensitivity-enhanced electron-holographic interferometry and thickness-measurement applications at atomic scale. Phys. Rev. Lett. **54**, 60 (1985)
7.4 S. Iijima, T. Ichihashi: Single shell carbon nanotubes of one nanometer diameter. Nature **363**, 603 (17 July 1993)
7.5 Q. Ru: Phase-shifting techniques in electron holography, in *Proc. Int'l Workshop on Electron Holography*, Knoxville, TN, ed. by A. Tonomura, L.F. Allard, G. Pozzi, D.C. Joy, Y.A. Ono (Elsevier, Amsterdam 1995) p. 69

7.6 X. Lin, V. Ravikumar, R.P. Rodrigues, N. Wilcox, V.P. Dravid: Electron holography in material science, in *Proc. Int'l Workshop on Electron Holography*, Knoxville, TN, ed. by A. Tonomura, L.F. Allard, G. Pozzi, D.C. Joy, Y.A. Ono (Elsevier, Amsterdam 1995) p.209

7.7 L.F. Allard, E. Völkl, S. Subramoney, R.S. Ruoff: Electron holography applied to the study of fullerence materials, in *Proc. Int'l Workshop on Electron Holography* (Knoxville, TN), ed. by A. Tonomura, L.F. Allard, G. Pozzi, D.C. Joy, Y.A. Ono (Elsevier, Amsterdam 1995) p.219

7.8 G. Lulli, P.G. Merli, A. Migliori, G. Mateucci, M. Stanghellini: Characterization of defects produced during self-annealing implantation of As in silicon. J. Appl. Phys. **68**, 2708 (1990)

7.9 A.K. Datye, D.S. Kalakkad, E. Völkl, L.F. Allard: Electron holography of heterogeneous catalysis, in *Proc. Int'l Workshop on Electron Holography* (Knoxville, TN), ed. by A. Tonomura, L.F. Allard, G. Pozzi, D.C. Joy, Y.A. Ono (Elsevier, Amsterdam 1995) p.199

7.10 T. Kawasaki, J. Endo, T. Matsuda, N. Osakabe, A. Tonomura: Application of holographic interference electron microscopy to the observation of biological specimens. J. Electron Microsc. **35**, 211 (1986)

7.11 H.L. Cox Jr., R.A. Bonham: Elastic electron scattering amplitudes for neutral atoms calculated using the partial wave method at 10, 40, 70 and 100 kV for $Z = 1$ to $Z = 54$. J. Chem. Phys. **47**, 2599 (1967)

7.12 N. Osakabe, J. Endo, T. Matsuda, A. Tonomura, A. Fukuhara: Observation of surface undulation due to single-atomic shear of a dislocation by reflection-electron holography. Phys. Rev. Lett. **62**, 2969 (1989)

7.13 P.G. Merli, G.F. Missiroli, G. Pozzi: p-n junction observations by interference electron microscopy. J. de Microscopie **21**, 11 (1974)

7.14 Yu.A. Kulyupin, S.A. Nepijko, N.N. Sedov, V.G. Shamonya: Use of interference microscopy to measure electric field distributions. Optik **52**, 101 (1978/79)

7.15 S. Frabboni, G. Matteucci, G. Pozzi, M. Vanzi: Electron holographic observation of the electrostatic field associated with thin reverse-biased p-n junctions. Phys. Rev. Lett. **55**, 2196 (1985)

7.16 B. Lau, G. Pozzi: Off axis electron micro-holography of magnetic domain walls. Optik **51**, 287 (1978)

7.17 H. Wahl, B. Lau: Theoretische Analyse des Verfahrens, die Feldvertailung in dünnen magnetischen Schichten durch lichtholographische Auswertung elektroneninterferenzmikroskopischer Aufnahmen zu veranschaulichen. Optik **54**, 27 (1979)

7.18 A. Tonomura, T. Matsuda, J. Endo, T. Arii, K. Mihama: Direct observation of fine structure of magnetic domain walls by electron holography. Phys. Rev. Lett. **44**, 1430 (1980)

7.19 T. Matsuda, A. Tonomura, R. Suzuki, J. Endo, N. Osakabe, H. Umezaki, H. Tanabe, Y. Sugita, H. Fujiwara: Observation of microscopic distribution of magnetic field by electron holography. J. Appl. Phys. **53**, 544 (1982)

7.20 A. Fukuhara, K. Shinagawa, A. Tonomura, H. Fujiwara: Electron holography and magnetic specimens. Phys. Rev. B **27**, 1839 (1983)

7.21 H. Koch, H. Lübbig (eds.): *Superconducting Devices and Their Applications*, Springer Proc. Phys., Vol.64 (Springer, Berlin, Heidelberg 1992)

7.22 M.S. Cohen: Wave-optical aspects of Lorentz microscopy. J. Appl. Phys. **38**, 4966 (1967)

7.23 A. Tonomura: The electron interference method for magnetization measurement of thin films. Jpn. J. Appl. Phys. 11, 493 (1972)

7.24 G. Pozzi, G.F. Missiroli: Interference electron microscopy of magnetic domains. J. Microscopie 18, 103 (1973)

7.25 A. Tonomura, T. Matsuda, H. Tanabe, N. Osakabe, J. Endo. A. Fukuhara, K. Shinagawa, H. Fujiwara: Electron holography technique for investigating thin ferromagnetic films. Phys. Rev. B 25, 6799 (1982)

7.26 E.E. Huber, D.O. Smith, J.B. Goodenough: Domain-wall structure in permalloy films. J. Appl. Phys. 29, 294 (1958)

7.27 T. Arii, S. Yatsuya, N. Wada, K. Mihama: Ferromagnetic domains in F.C.C. cobalt fine particles prepared by gas-evaporation technique. Proc. 5th Int'l Conf. High Voltage Electrton Microscopy (Kyoto 1977) (Jpn. Soc. Electron Microscopy, Kyoto 1977) pp. 203-206

7.28 N. Osakabe, K. Yoshida, Y. Horiuchi, T. Matsuda, H. Tanabe, T. Okuwaki, J. Endo, H. Fujiwara, A. Tonomura: Observation of recorded magnetization pattern by electron holography. Appl. Phys. Lett. 42, 746 (1983)

7.29 K. Yoshida, T. Okuwaki, N. Osakabe, H. Tanabe, Y. Horiuchi, T. Matsuda, K. Shinagawa, A. Tonomura, H. Fujiwara: Observation of recorded magnetization patterns by electron holography. IEEE Trans. Magn. MAG-19, 1600 (1983)

7.30 S. Iwasaki, T. Nakamura: An analysis for the magnetization mode for high density magnetic recording, IEEE Trans. MAG-13, 1272 (1977)

7.31 A. Tonomura, T. Matsuda, J. Endo, T. Arii, K. Mihama: Holographic interference electron microscopy for determining specimen magnetic structure and thickness distribution. Phys. Rev. B 34, 3397 (1986)

7.32 G. Lai, T. Hirayama, K. Ishizuka, T. Tanji, A. Tonomura: Three-dimensional reconstruction of electrical-potential distribution in electron-holographic interferometry. Appl. Opt. 33, 829 (1994)

7.33 G. Lai, T. Hirayama, K. Ishizuka, T. Tanji, Tonomura: Three-dimensional reconstruction of electric-potential distribution in electron-holographic interferometry. Appl. Opt. 33, 829 (1994)

7.34 U. Essmann, H. Träuble: The direct observation of individual flux lines in type II superconductors. Phys. Lett. A 24, 526 (1967)

7.35 J. Mannhart, J. Bosch, R.P. Huebener: Elementary pinning forces measured using low temperature scanning electron microscopy. Phys. Lett. A 122, 439 (1987)

7.36 H.F. Hess, R.B. Robinson, R.C. Dynes, J.M. Valles, Jr., J.V. Waszczak: Scanning-tunneling-microscope observation of the Abrikosov flux lattice and the density of states near and inside a fluxoid. Phys. Rev. Lett. 62, 214 (1989)

7.37 A.M. Chang, H.D. Hallen, L. Harriott, H.F. Hess, H.L. Kao, J. Kwo, R.E. Müller, R. Wolfe, J. van der Ziel, T.Y. Chang: Scanning Hall probe microscopy. Appl. Phys. Lett. 61, 1974 (1992)

7.38 A. Oral, S.J. Bending: Real-time scanning Hall probe microscopy. Appl. Phys. Lett. 69, 1324 (1996)

7.39 L.N. Vu, M.S. Wistrom, D.J. Van Harlingen: Imaging of magnetic vortices in superconducting networks and clusters by scanning SQUID microscopy. Appl. Phys. Lett. 63, 1693 (1993)

7.40 A. Moser, H.J. Hug, I. Parashikov, B. Stiefel, O. Fritz, H. Thomas, A. Baratoff, H.-J. Güntherodt: Observation of single vortices condensed into a vortex-glass phase by magnetic force microscopy: Phys. Rev. Lett. 74, 1847 (1995)

7.41 T. Matsuda, H. Hasegawa, M. Igarashi, T. Kobayashi, M. Naito, H. Kajiyama, J. Endo, N. Osakabe, A. Tonomura, R. Aoki: Magnetic field observation of a single flux quantum by electron-holographic interferometry. Phys. Rev. Lett. **62**, 2519 (1989)

7.42 S. Hasegawa, T. Matsuda, J. Endo, N. Osakabe, M. Igarashi, T. Kobayashi, M. Naito, A. Tonomura, R. Aoki: Magnetic-flux quanta in superconducting thin films observed by electron holography and digital phase analysis. Phys. Rev. B **43**, 7631 (1991)

7.43 J.M. Kosterlitz, D.D.J. Thouless: Ordering, metastability and phase transitions in two-dimensional systems. J. Phys. **6**, 1181 (1973)

7.44 B.I. Halperin, D.R. Nelson: Resistive transition in superconducting films. J. Low Temp. Phys. **36**, 599 (1979)

7.45 T. Matsuda, A. Fukuhara, T. Yoshida, S. Hasegawa, A. Tonomura, Q. Ru: Computer reconstruction from electron holograms and observation of fluxon dynamics. Phys. Rev. Lett. **66**, 457 (1991)

7.46 Q. Ru, T. Matsuda, A. Fukuhara, A. Tonomura: Digital extraction of the magnetic-flux distribution from an electron interferogram. J. Opt. Soc. Am. **8**, 1739 (1991)

7.47 T. Yoshida, T. Matsuda, A. Tonomura: Electron holography observation of flux-line dynamics, Proc. 50th Meeting of Electron Microscopy Society of America, Boston 1992, ed. by G.W. Bailey, J. Bentley, J.A. Small (San Francisco Press, San Francisco, 1992) pp. 68-69

7.48 G.S. Park, C.E. Cunningham, B. Cabrera, M.E. Huber: Vortex pinning force in a superconducting niobium strip. Phys. Rev. Lett. **68**, 1920 (1992)

7.49 O.B. Hyun, D.K. Finnemore, L. Scharztkopf, J.R. Clem: Elementary pinning force for a superconducting vortex. Phys. Rev. Lett. **58**, 599 (1987)

7.50 H. Yoshioka: On the electron diffraction by flux lines. J. Phys. Soc. Jpn. **21**, 948 (1960)

7.51 M.J. Goringe, J.P. Jakubovics: Electron diffraction from periodic magnetic fields. Phil. Mag. **15**, 393 (1967)

7.52 J.P. Guigay, A. Bourret: Calcul des franges de defocalisation d'une ligne de vortex, en microscopie electronique. C.R. Acad. Sci. (Paris) **264**, 1389 (1967)

7.53 D. Wohlleben: Diffraction effects in Lorentz microscopy. J. Appl. Phys. **38**, 3341 (1967)

7.54 C. Colliex, B. Jouffrey, M. Kleman: Sur les possibilities d'observation de sligne de vortex en microscopie electronique par transmission. Acta Cryst. A **24**, 692 (1968)

7.55 C. Capiluppi, G. Pozzi, U. Valdrè: On the possibility of observing fuxons by transmission electron microscopy. Phil. Mag. **26**, 865 (1972)

7.56 A. Migliori, G. Pozzi: Computer simulation of electron holographic contour maps of superconducting flux lines. Ultramicroscopy **41**, 169 (1992)

7.57 A. Migliori, G. Pozzi, A. Tonomura: Computer simulation of electron holographic contour maps of superconducting flux lines II. The case of tilted specimen. Ultramicroscopy **49**, 87 (1993)

7.58 K. Harada, T. Matsuda, J. Bonevich, M. Igarashi, S. Kondo, G. Pozzi, U. Kawabe, A. Tonomura: Real-time observation of vortex lattices in a superconductor by electron microscopy. Nature **360**, 51 (5 November 1992)

7.59 J.E. Bonevich, K. Harada, T. Matsuda, H. Kasai, T. Yoshida, G. Pozzi, A. Tonomura: Electron holography observation of vortex lattices in a superconductor. Phys. Rev. Lett. **70**, 2952 (1993)

7.60 T. Matsuda, K. Harada, H. Kasai, O. Kamimura, A. Tonomura: Observation of dynamic interaction of vortices with pinning centers by Lorentz microscopy. Science **271**, 1393 (8 March 1996)

7.61 G.W. Crabtree, D.R Nelson: Vortex physics in high-temperature superconductors. Phys. Today **50**, 38 (April 1997)

7.62 C. Reichhardt, J. Groth, C.J. Olson, S.B. Field, F. Nori: Spatiotemporal dynamics and plastic flow of vortices in superconductors with periodic arrays of pinning sites. Phys. Rev. B **54**, 16108 (1996)

7.63 C. Reichahrdt, C.J. Olson, F. Nori: States in Dynamic phases of vortices in superconductors with periodic pinning arrays. Phys. Rev. Lett. **78**, 2648 (1997)

7.64 K. Harada, O. Kamimura, H. Kasai, T. Mastuda, A. Tonomura: Direct observation of vortex dynamics in superconducting films with regular arrays. Science **274**, 1167 (15 November 1996)

7.65 K. Harada, H. Kasai, T. Matsuda, M. Yamasaki, A. Tonomura: Direct observation of interaction of vortices and antivortices in a superconductor by Lorentz microscopy. J. Electron Microsc. **46**, 227 (1997)

7.66 K. Harada, H. Kasai, J.E. Bonevich, T. Yoshida, A. Tonomura: Vortex configuration and dynamics in $Bi_2 Sr_{1.8} CaCuO_x$ thin film by Lorentz microscopy. Phys. Rev. Lett. **71**, 3371 (1993)

7.67 For example, see D.J. Bishop, P.L. Gammel, D.A. Huse, C.A. Murray: Magnetic flux-line lattices and vortices in the copper oxide superconductors. Science **255**, 165 (10 January 1992)

7.68 R.N. Kleiman, P.L. Gammel, L.F. Schneemeyer, J.V. Waszczak, D.J. Bishop: Kleiman et al. Reply. Phys. Rev. Lett. **62**, 2331 (1989)

7.69 A. Tonomura, H. Kasai, O. Kamimura, T. Matsuda, K. Harada, J. Shimoyama, K. Kisho, K. Kitazawa: Motion of vortices in superconductors. Nature **397**, 308 (28 January 1999)

Chapter 8

8.1 O. Scherzer: Über einige Fehler von elektronen Linsen. Z. Phys. **2**, 593 (1936)

8.2 F. Zernike: Phase contrast, a new method for the observation of transparent objects. Physica **9**, 686 (1942)

8.3 F. Zernike: Phase contrast, a new method for the observation of transparent objects, Part II. Physica **9**, 974 (1942)

8.4 O. Scherzer: The theoretical resolution limit of the electron microscope. J. Appl. Phys. **20**, 20 (1949)

8.5 K.-J. Hanszen: The optical transfer theory of the electron microscope: Fundamental principles and applications. In *Advanced Optical Electron Microscopy*, ed. by R. Barer, V.E. Cosslett (Academic, London 1971) Vol.4, pp.1-84

8.6 A. Tonomura, T. Matsuda, J. Endo: Spherical-aberration correction of electron lens by holography. Jpn. J. Appl. Phys. **18**, 1373 (1979)

8.7 S. Ino: Epitaxial growth of metals on rocksalt faces cleaved in vacuum. II. Orientation and structure of gold particles formed in ultrahigh vacuum. J. Phys. Soc. Jpn. **21**, 346 (1966)

8.8 K. Mihama, Y. Yasuda: Initial stage of epitaxial growth of evaporated gold films on sodium chloride. J. Phys. Soc. Jpn. **21**, 1161 (1966)

8.9 H. Lichte: Electron holography approaching atomic resolution. Ultramicroscopy **20**, 293 (1986)

8.10 T. Kawasaki, Q. Ru, T. Matsuda, Y. Bando, A. Tonomura: High resolution holography observation of H-Nb_2O_5. Jpn. J. Appl. Phys. **30**, L1830 (1991)

8.11 A. Orchowski, W.D. Rau, H. Lichte: Electron holography surmounts resolution limit of electron microscopy. Phys. Rev. Lett. **74**, 399 (1995)

8.12 G. Lang, M. Lehmann, D.J. Smith, M.R. McCartney, H. Lichte: High resolution electron holography of CdTe and ZnTe. Proc. MRS'97 (Fall 1998 Meeting) to be published

8.13 W. Coene, G. Janssen, M.O. de Beeck, D.V. Dyck: Phase retieval through focus variation for ultra-resolution in field-emission transmission electron microscopy: Phys. Rev. Lett. **69**, 3743 (1992)

Subject Index

Aberration 1
- compensation 44
- correction 44
Accelerating voltage 10
Acceleration tube 31
Aharonov Y. 50
Aharonov-Bohm (AB) effect 50
Amplified interference micrograph 39
Amplitude object 136
Amplitude-division beam splitter 25
Amplitude transmittance 2
Antifluxon 127
Atomic step 80

Beryllium 80
Biprism fringes 26
Biprism interference pattern 16
Bitter figure 89
Bitter method 104
Bocchieri P. 61
Boersch H. 20
Boersch effect 16
Bohm D. 50
Bragg-reflected beam 27
Bragg-reflected image 140
Bragg reflection 27
Brightness 10
Bubble memory 95

Carbon nanotube 81
Carrier fringe 27
Cathode 17
Chambers R. G. 60
Chromatic aberration 10
Cobalt 48
Coherence
- length 16
- property 11
Concave electron lens 12
Condenser lens 10
Conjugate image 3

Contact potential 13
Contamination 47
Contour fringe 1
Contour map 34
Contrast transfer 136
Cooper pair 86
Coulomb gauge 61
Cross-tie wall 89
Current density 19
Curvature of surface 59

Defocusing 132
- distance 134
Delta function 64
Dennis Gabor 1
Dislocation core 83
Double-slit experiment 53
Dynamic observation 102

Ehrenberg W. 63
Electric-field distribution 84
Electro-kinetic momentum 55
Electromagnetic induction 55
Electromagnetic potential 50
Electron beam 1
Electron biprism 13
Electron gun 10
Electron interferometer 32
Electron lens 12
Electron microscopy 10
Electron optics 10
Electron hologram 29
Electron-holographic interferometry 74
Electronic state 55
Equipotential line 84

Faraday M. 55
Fermat's principle 63
Ferritine 82
Ferromagnetic film 1
Ferromagnetic particle 87

159

Fiber bundle 56
Field emission 1
- electron beam 1
- electron gun 11
- tip 29
Flux line 111
Flux pinning 116
Flux quantization 77
Flux quantum 107
Fluxon 77
Fluxon lattice 119
Fourier-transform hologram 28
Fourier-transform holography 28
Fourier-transform method 39
Fraunhofer condition 22
Fraunhofer in-line holography 5
Fresnel fringe 20
Fringe-scanning method 39

GaAs 83
Gabor D. 1
Gauge field 50
Gauge principle 50
Gauge transformation 62
Gaussian image 44
Gold 14

Hairpin-type cathode 16
Half-spacing fringe 27
Hanszen K.-J. 24
Hibi T. 17
High-resolution hologram 27
High-resolution microscopy 132
High-T_c superconductor 103
Holography 1
- electron microscopy 30
Hydrodynamical formulation 62

Iijima S. 81
Illumination angle 10
Image resolution 5
In-line holography 5
In-line projection holography 20
In-line transmissin holography 22
In-plane magnetic recording 94
Inner potential 13
Interference electron microscope 13
Interference microscopy 34
Interferogram 14
Interferometer 13
Intermediate state 107
Iron whisker 60

Kosterlitz-Thouless theory 107
Kuper C.G. 65

LaB_6 11
Laser 1
Laterial magnification 9
Lattice fringe 27
Lattice image 28
Lead 105
Leakage flux 67
Leith E.N. 1
Lichte H. 27
Liouvill's theorem 19
Lithography 66
Local interaction principle 54
Loinger A. 61
Longitudinal coherence 16
Longitudinal magnification 8
Lorentz force 54, 60, 61, 76, 77, 89, 100
Lorentz micrograph 62
Lorentz microscopy 87
Low-angle electron diffraction 48

Mach-Zehnder interferometer 37
Magnesium oxide 14
Magnetic-domain structure 1
Magnetic-domain wall 87
Magnetic electron lens 12
Magnetic fluxon 1
Magnetic line of force 1
Magnetic recording 93
Magnetic tape 93
Magnetization 49
Magnetostatic energy 89
Manifold 58
Matching effect 124
Maxwell J.C. 55
Maxwell's equation 55
Meissner effect 70
Micro-area electron diffraction 47
Mills R.L. 56
Missiroli G. 87
Molybdenite 80
Multi-valued wave function 69
Multiple-beam interferometry 80
Multiply-twinned particle 137
Möllenstedt G. 13

Néel's wall 88
Nickel 90
Niobium oxide 142
Non-abelian gauge field 50

Non-integrable phase factor 56
Non-Stokesian vector potential 62
Nonexistence of the Aharonov-Bohm effect 61
Nonlinear hologram 37

Object wave 3
Off-axis electron hologram 29
Off-axis Fresnel hologram 28
Off-axis holography 5
Off-axis "image" holography 25
Optical diffraction pattern 142
Overfocusing 135

p-n junction 84
Parallel transport 56
Paraxial ray equations 12
Pauli W. 69
Perpendicular magnetic recording 94
Phase amplification 37
Phase
- contrast 82
- difference 36
- factor 56
  grating 134
- image 27
- object 24
- -amplification technique 30
- -amplified hologram 80
- -amplified interference microscopy 36
- -contrast transfer function 134
- -shifting method 39
Piezoelectric transducer 39
Pinning center 103
Plastic flow 124
Point resolution 46
Pointed cathode 17
Pointed filament 11
Pozzi G. 87
Principal connection 59
Profile mode of holography 103

Radiation damage 47
Real-time observation 39
Reflection mode of holography 83
Refractive index 12
Resolution 1

Salam A. 56
Scanning electron microscope 104
Scanning tunneling microscope 104
Scattering amplitude 3

Scherzer O. 132
Schrödinger E. 69
Schrödinger equation 50
Screw dislocation 83
Siday R.W. 63
Side band 34
Simply connected region 58
Single-atom field emitter 20
Single-atom tip 20
Single-valued wave function 69
Solenoid 51
Spatial coherence 16
Spatial coherence length 17
Spherical aberration 12, 46
Spherical aberration correction 24
SQUID 86
Stoke's theorem 62
Structure image 27
Superconducting layer 70
Superconducting toroid 74
Superconductivity 103
Surface topography 83

Takabayasi T. 64
Temporal coherence 16
Theory of fiber bundles 56
Theory of gauge fields 50
Thermal electron beam 17
Thermal emission 11
Thermally excited fluxon 108
Thickness measurement 78
Thouless D.D.J. 107
Time-reversal operation 96
Toroidal 66
Toroidal solenoid 66
Toroidal superconductor 73
Transfer function 24
Transmission electron microscope 10
Transmission mode of holography 83
Transmitted wave 20
Transparent toroidal magnet 66
Transverse coherence 16
Twin image 5
Twyman-Green interferometer 39
Type-I superconductor 103
Type-II superconductor 103

Underfocusing 135
Upatnieks J. 1
Utiyama R. 56

Vector potential 50

Void 81
Vortex 103

Wave front 1
- aberration 45
- division beam splitter 27
Wave function 1
Wave number 8
Wave packet 16
Wavelength 1

Weinberg S. 56
Work function 17
Wu T.T. 56

Yang C.N. 56
Young's experiment 53

Zinc oxide 21
Zone plate 4

Springer Series in Optical Sciences

Editorial Board: A. L. Schawlow A. E. Siegman T. Tamir

Managing Editor: H. K. V. Lotsch

1 **Solid-State Laser Engineering**
  By W. Koechner 5th Edition

2 **Table of Laser Lines in Gases and Vapors**
  By R. Beck, W. Englisch, and K. Gurs
  3rd Edition

3 **Tunable Lasers and Applications**
  Editors: A. Mooradian, T. Jaeger,
  and P. Stokseth

4 **Nonlinear Laser Spectroscopy**
  By V. S. Letokhov and V. P. Chebotayev
  3nd Edition

5 **Optics and Lasers** Including Fibers
  and Optical Waveguides By M. Young
  4th Edition (available as a textbook)

6 **Photoelectron Statistics**
  With Applications to Spectroscopy
  and Optical Communication By B. Saleh

7 **Laser Spectroscopy III**
  Editors: J. L. Hall and J. L. Carlsten

8 **Frontiers in Visual Science**
  Editors: S. J. Cool and E. J. Smith III

9 **High-Power Lasers and Applications**
  Editors: K.-L. Kompa and H. Walther

10 **Detection of Optical and Infrared Radiation**
   By R. H. Kingston

11 **Matrix Theory of Photoelasticity**
   By P. S . Theocaris and E. E. Gdoutos

12 **The Monte Carlo Method in Atmospheric Optics**
   By G. I. Marchuk, G. A. Mikhailov,
   M. A. Nazaraliev, R. A. Darbinian, B. A. Kargin,
   and B. S. Elepov

13 **Physiological Optics**
   By Y. Le Grand and S. G. El Hage

14 **Laser Crystals** Physics and Properties
   By A. A. Kaminskii 2nd Edition

15 **X-Ray Spectroscopy** By B. K. Agarwal
   2nd Edition

16 **Holographic Interferometry**
   From the Scope of Deformation Analysis
   of Opaque Bodies
   By W. Schumann and M. Dubas

17 **Nonlinear Optics of Free Atoms and Molecules**
   By D. C. Hanna, M. A. Yuratich, and D. Cotter

18 **Holography in Medicine and Biology**
   Editor: G. von Bally

19 **Color Theory and Its Application in Art
   and Design** By G. A. Agoston 2nd Edition

20 **Interferometry by Holography**
   By Yu. I. Ostrovsky, M. M. Butusov,
   and G. V. Ostrovskaya

21 **Laser Spectroscopy IV**
   Editors: H. Walther and K. W. Rothe

22 **Lasers in Photomedicine and Photobiology**
   Editors: R. Pratesi and C. A. Sacchi

23 **Vertebrate Photoreceptor Optics**
   Editors: J. M. Enoch and F. L. Tobey, Jr.

24 **Optical Fiber Systems and Their Components**
   An Introduction By A. B. Sharma,
   S. J. Halme, and M. M. Butusov

25 **High Peak Power Nd: Glass Laser Systems**
   By D. C. Brown

26 **Lasers and Applications**
   Editors: W. O. N. Guimaraes, C. T. Lin,
   and A. Mooradian

27 **Color Measurement**
   Theme and Variations
   By D. L. MacAdam 2nd Edition

28 **Modular Opticsl Design**
   By O. N. Stavroudis

29 **Inverse Problems of Lidar Sensing of the
   Atmosphere** By V. E. Zuev and I. E. Naats

30 **Laser Spectroscopy V**
   Editors: A. R. W. McKellar, T. Oka,
   and B. P. Stoicheff

31 **Optics in Biomedical Sciences**
   Editors: G. von Bally and P. Greguss

32 **Fiber-Optic Rotation Sensors**
   and Related Technologies
   Editors: S. Ezekiel and H. J. Arditty

33 **Integrated Optics:** Theory and Technology
   By R. G. Hunsperger
   3th Edition (available as a textbook)

34 **The High-Power Iodine Laser**
   By G. Brederlow, E. Fill, and K. J. Witte

35 **Engineering Optics**
   By K. Iizuka 2nd Edition

36 **Transmission Electron Microscopy**
   Physics of Image Formation and Microanalysis
   By L. Reimer 4th Edition

37 **Opto-Acoustic Molecular Spectroscopy**
   By V. S. Letokhov and V. P. Zharov

38 **Photon Correlation Techniques**
   Editor: E. O. Schulz-DuBois

39 **Optical and Laser Remote Sensing**
   Editors: D. K. Killinger and A. Mooradian

40 **Laser Spectroscopy VI**
   Editors: H. P. Weber and W. Lüthy

41 **Advances in Diagnostic Visual Optics**
   Editors: G. M. Breinin and I. M. Siegel

Printing: Saladruck, Berlin
Binding: H. Stürtz AG, Würzburg